Cambridge Elements ≡

Elements in Flexible and Large-Area Electronics
edited by
Ravinder Dahiya
University of Glasgow
Luigi G. Occhipinti
University of Cambridge

HYBRID SYSTEMS-IN-FOIL

Mourad Elsobky
IMS CHIPS
Joachim N. Burghartz
IMS CHIPS

CAMBRIDGE
UNIVERSITY PRESS

CAMBRIDGE
UNIVERSITY PRESS

University Printing House, Cambridge CB2 8BS, United Kingdom

One Liberty Plaza, 20th Floor, New York, NY 10006, USA

477 Williamstown Road, Port Melbourne, VIC 3207, Australia

314–321, 3rd Floor, Plot 3, Splendor Forum, Jasola District Centre, New Delhi – 110025, India

103 Penang Road, #05-06/07, Visioncrest Commercial, Singapore 238467

Cambridge University Press is part of the University of Cambridge.

It furthers the University's mission by disseminating knowledge in the pursuit of education, learning, and research at the highest international levels of excellence.

www.cambridge.org
Information on this title: www.cambridge.org/9781108984744
DOI: 10.1017/9781108985116

First published 2021

A catalogue record for this publication is available from the British Library.

ISBN 978-1-108-98474-4 Paperback
ISSN 2398-4015 (online)
ISSN 2514-3840 (print)

Hybrid Systems-in-Foil

Elements in Flexible and Large-Area Electronics

DOI: 10.1017/9781108985116
First published online: September 2021

Mourad Elsobky
IMS CHIPS

Joachim N. Burghartz
IMS CHIPS

Author for correspondence: Joachim N. Burghartz, burghartz@ims-chip.de

Abstract: Hybrid Systems-in-Foil (HySiF) is a concept that extends the potential of conventional More-than-More Systems-in/on-Package (SiPs and SoPs) to the flexible electronics world. In HySiF, an economical implementation of flexible electronic systems is possible by integrating a minimum number of embedded silicon chips and a maximum number of on-foil components. Here, the complementary characteristics of CMOS SoCs and larger-area organic and printed electronics are combined in a HySiF-compatible polymeric substrate. Within the HySiF scope, the fabrication process steps and the integration design rules together with all the accompanying boundary conditions concerning material compatibility, surface properties, and thermal budget are defined. This Element serves as an introduction to the HySiF concept. A summary of recent ultrathin chip fabrication and flexible packaging techniques is presented. Several bendable electronic components are discussed, demonstrating the benefits of HySiF. Finally, prototypes of flexible wireless sensor systems that adopt the HySiF concept are demonstrated.

Keywords: Hybrid System-in-Foil, flexible electronics, ultrathin chip, ChipFilm Patch, electronic skin

ISBNs: 9781108984744 (PB), 9781108985116 (OC)
ISSNs: 2398-4015 (online), 2514-3840 (print)

Contents

1 Flexible Electronics and Hybrid Systems-in-Foil

1.1 Motivation

Silicon device miniaturization and very-large-scale integration (VLSI) have followed the well-known Moore's law for the last 50 years.[1] Integrated digital functions and information processing have downscaled steadily in what is known as the "More Moore" (MM) trend. Alternatively, real-world sensory interactions and analog functional diversification have indirectly benefited from Moore's law, but they do not necessarily scale in size or cost, which gave rise to a new growth trend known as "More than Moore" (MtM). Figure 1.1a illustrates the timeline of CMOS technology scaling and highlights topics that drive the digital and non-digital integrated functionality.

Complex digital integrated circuits (ICs) were developed on a silicon die in a so-called system-on-chip (SoC) approach. For implementing more system functionality and due to limited access to custom IC foundries, multiple chips were jointly assembled for achieving a higher abstraction level, arranged in a so-called multi-chip module (MCM). The steady scaling of devices in digital ICs is accompanied by continuous improvement of miniaturized passive and active components such as capacitors, inductors, micro-electromechanical systems (MEMS), and analog circuitry, which, when combined with SoCs, give rise to a new level of integration known as system-in-package (SiP) and later system-on-package (SoP).

Reports cite the world's largest IC foundry Taiwan Semiconductor Manufacturing Company (TSMC) as having said, "Moore's Law is not dead, it's not slowing down, it's not even sick."[2] However, several physical, material, power-thermal, technological, and economical challenges are facing the continuation of device miniaturization, and reports are already speculating on the imminent end or slowing down of Moore's law.[3], [4]

Other materials such as gallium arsenide (GaAs), gallium nitride (GaN), and small-molecule organic semiconductors have found their application fields away from scaled silicon technology. Nonetheless, non-silicon semiconductor manufacturing benefits from the success story of silicon integration and miniaturization. Figure 1.1b shows different semiconductor materials and their operating frequency versus power-handling ranges. Note the optimal place occupied by the irreplaceable high-performance silicon semiconductor industry, which has led to the modern digital age of the Internet, computing, and portable electronics.

1.2 Flexible Electronics

Flexible electronics is considered one of the main enablers of the Internet of Things (IoT), as it introduces smartness to every *Thing* in our daily life

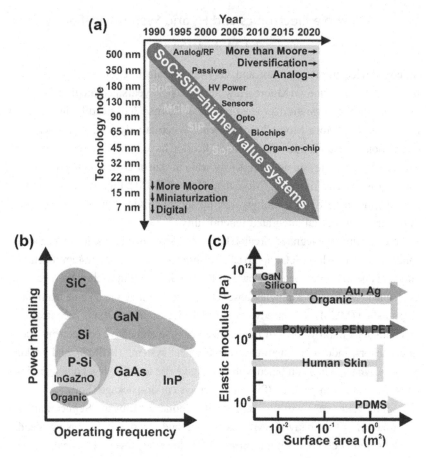

Figure 1.1 (a) The combined need for digital and non-digital functionalities in an integrated system is translated as a dual trend in miniaturization of the digital functions ("More Moore") and functional diversification ("More than Moore"). [5] (b) Power-handling capability of different semiconductor materials illustrated against the operating frequency range. (c) Elasticity of frequently used materials in flexible and printed electronics and its potential surface area coverage.

regardless of its form factor and surface properties. Flexible, printable, and organic electronics, when combined with new materials and advanced fabrication processes, offer unique characteristics such as mechanical flexibility, thin-form factor, large area scaling feasibility, and adaptability to irregular surfaces. However, most flexible electronic components are currently either fabricated as stand-alone components or combined with bulky sensor readout and/or wireless communication modules assembled to the surface of a flexible substrate.[6], [7], [8] Other flexible electronic systems benefit from only one technology (e.g., integrated temperature sensor and analog-to-digital converter [ADC] using amorphous Indium-Gallium-Zinc-Oxide [InGaZnO] thin-film transistors [TFTs]).[9]

Figure 1.1c shows the elasticity of frequently used materials in flexible and printed electronics and its potential surface area coverage. Thin-film organic, small-molecule, and inorganic poly-Si semiconductor materials are inherently flexible and applied to low-speed and large-area electronic applications. Due to the strengthened requirement on material purity and uniformity, single-crystalline silicon has a limited surface area coverage governed by the maximum available wafer size. Ultrathin chips (UTCs) are a promising candidate for high-performance bendable electronic applications with embedded intelligence. UTCs can be considered the extension of mature silicon technology to flexible electronics. However, a cost-effective and reliable UTC flexible packaging solution is still in development. Note that GaN and GaAs on thin silicon have extended the operating frequency of flexible and stretchable microwave devices and circuits reaching 10 GHz and beyond.[10], [11] Substrates based on polyimides, polyethylene naphthalate (PEN), and polydimethylsiloxane (PDMS) are potential platforms for flexible systems integration complementing the conventional rigid printed circuit boards (PCBs).

1.2.1 Hybrid Systems-in-Foil Definition

Hybrid Systems-in-Foil (HySiF) is a concept that extends the potential of conventional MtM SiPs and SoPs to the flexible electronics world. HySiF targets an economical implementation of flexible electronic systems by integrating a minimum number of embedded silicon chips and a maximum number of on-foil components.[12], [13] The complementary characteristics of CMOS SoCs and larger-area organic and printed electronics are combined in a HySiF-compatible polymeric substrate. Within the HySiF scope, the fabrication process steps and design rules for integrating such flexible electronic components with all the accompanying boundary conditions concerning material compatibility, surface properties, and thermal budget are defined.

In many electronic systems (e.g., data converters and microprocessors), standard criteria for performance evaluation are designed to reflect the trade-offs between operational parameters such as power consumption, speed, and accuracy. Walden and Schreier figures of merit (FoM) are known values to evaluate data converters' performance. In this context, a new figure of merit (FoM_{Flex}) for flexible smart electronic systems (i.e., bendable and/or stretchable electronic systems with digital outputs) includes key features such as mechanical flexibility, low-power operation, accuracy, and processing speed. This newly introduced FoM is defined as follows (Eq. 1.1):

$$FoM_{Flex}(Pa.J) = \frac{E\ (Pa) \times P\ (W)}{F_{sync}\ (s^{-1}) \times SNR},\tag{1.1}$$

where E is Young's (or the Elastic) modulus of the complete flexible system or substrate, P is the power consumption, F_{sync} is the maximum data rate supported by the flexible system, and SNR is the signal-to-noise ratio of the integrated front end. More elasticity, higher speed, and SNR at lower power results in better FoM. As shown in Figure 1.3, silicon-based UTCs outperform organic and inorganic TFT technologies. However, large area coverage, low-temperature processing, and shorter manufacturing time are the main advantages for organic and inorganic TFT technologies and are challenging parameters to be evaluated using a standardized FoM.

For purely digital IP blocks, such as microprocessors (μP), microcontrollers (μC), and near-field communication (NFC), the maximum data rate supported by the flexible system in samples/second and SNR of 2 are used. Figure 1.2 depicts the proposed FoM$_{Flex}$ against system energy consumption of selected flexible silicon, inorganic, and organic TFT smart systems. Table 1.1 provides a detailed comparison of the specifications of the blocks represented in Figure 1.2.

1.2.2 Device Integration and Interconnect Technologies

For flexible systems exploiting metallic foil substrates, wire interconnects are already available, and their properties are coupled to the substrate properties. In other systems that have insulating substrates, metallic films are deposited to electrically connect different electronic components. Conventional CMOS-compatible lithography could be used to trade off smaller area coverage and lower throughput with fine-pitch interconnects. Fortunately, the thinner the material, the more flexible it becomes. This is true in the case of thin-film metals, such as gold (Au), silver (Ag), and aluminum (Al), which normally are used to manufacture flexible electronic components. However, the thinner the metal, the higher the sheet resistance. As an example of this trade-off, multiple overpasses are performed during metal inkjet printing to tailor the sheet resistance to that of a printed single metal layer. Long wire interconnects that are often used in large-area flexible electronics affect signal integrity, as they limit the precision and speed of the embedded systems by adding line delays (higher parasitic RLC).

Coplanar circuit interconnects are usually used in flexible sensor systems. Since metals have higher Young's moduli compared to polymers, the density

Table 1.1 Comparison of selected smart flexible μCs, NFC, and sensors from literature

Pub.	Year	IP	Tech.	Substrate (E)	Supply (V)	Power (mW)	Fsync (Hz)	ENOB	FoM_F(Pa. J)
ISSCC [20]	2011	μC	Organic	Plastic (5GPa)*	10	0.092	6	1	38300
EMPC [83]	2011	μC	UTCP	PI (45GPa)*	2.2	1.43*	400k	1	80.4
EMPC [83]	2011	ADC-SAR	UTCP	PI (45GPa)*	2.2	0.726*	250k	11	0.0638
ISSCC [171]	2013	ADC-VCO	Organic	Plastic (5GPa)*	20	0.048	66	6	56.8
ISSCC [24]	2016	Stress sensor	ChipFilm	LCP (15GPa)	5	34.5	156k	10	3.24
ISSCC [144]	2017	ADC-ADSM	InGaZnO	PI (5GPa)	20	2	300	6.4	395
ISSCC [21]	2017	NFC	InGaZnO	Plastic (5GPa)*	5	7.5	106k	1	177
ELL [22]	2018	Strain gauge	CFP	PI (15GPa)	5	15	52k	10	4.23
JSSC [9]	2018	Temperature sensor	InGaZnO	PI (5GPa)*	30	245	400	5.7	58900
FLEPS [23]	2018	μC, Apollo	CFP	PI (15GPa)	2	3.74	922k	1	30.5
FLEPS [23]	2018	ADC-SAR, Apollo	CFP	PI (15GPa)	2	0.46	115k	8.7	0.15
Sensors J. [85]	2019	NFC, EM	CFP	PI (15GPa)	2	2	848k	1	17.7

* If E is not mentioned or ENOB and Power are not measured, values are assumed based on corresponding datasheet values

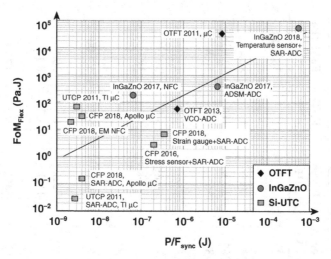

Figure 1.2 The proposed Figure of Merit (FoM$_{Flex}$) of Eq. 1.1 plotted vs. energy. Selected smart flexible electronic systems from the literature are divided into silicon UTCs, inorganic InGaZnO TFT, and organic TFT technologies. Note: flexible systems based on UTCs achieve better FoM$_F$ relative to organic and inorganic TFT technologies. Table 1.1 lists all data points that are plotted in this figure.

and layout of the circuit interconnects impact the freestanding warpage of the HySiF. In order to improve the elasticity of interconnects, various layout techniques have been proposed, such as zigzag and serpentine designs,[14] but this is usually done at the expense of longer wires. To this end, intrinsically elastic interconnects are reported using amorphous metals,[15] carbon nanotube (CNT),[16] or even transparent silver nanowires (Ag NW).[17], [18] However, the conductivity of conventional thin-film metals is still superior. Besides, the neutral plane of stress within the HySiF foil can be leveraged to improve the micro-mechanical reliability of critical components and circuit interconnects in particular.[19]

Screen, gravure, offset, and inkjet printing are well-known liquid printing techniques for manufacturing flexible and organic electronic components on metallic or polymeric substrates. These printing techniques are extended for mass production and to cover larger substrate areas in what is known as roll-to-roll (R2R) manufacturing. R2R processing is currently used for manufacturing simple flexible electronic products such as light-emitting diodes (LEDs) and e-ink displays for smart cards. The full potential of R2R technology can be unlocked by integrating high-performance electronic components such as microcontrollers or wireless communication modules.

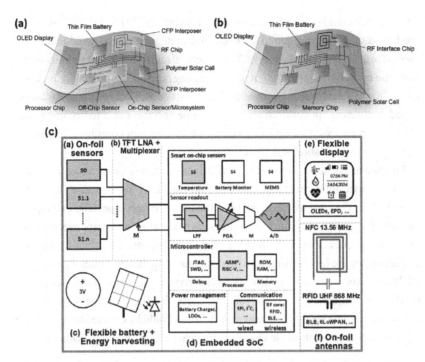

Figure 1.3 HySiF approaches in which UTCs along with on-foil flexible electronic components are combined. (a) ChipFilm Patch (CFP) technology is used as an interposer package in another flexible system.[12] (b) CFP itself is used as an integration platform for HySiF-compatible flexible components in SoP arrangement.[11] (c) Block diagram for a generic model of a smart sensor system reflecting the HySiF concept.

In [20], a simple microcontroller using inorganic TFTs is fabricated on a flexible substrate. In addition, NFC plastic circuitry is implemented using the same inorganic metal-oxide InGaZnO TFTs.[21] Other attempts use HySiF approaches such as in [22], [23], which exploit UTCs embedded in flexible packages as an interposer in another flexible motherboard, as illustrated in Figure 1.3a. In [24], [25], the UTC flexible package itself is used as the platform for integrating other flexible electronic components, as shown in Figure 1.3b. In a possible R2R manufacturing scenario, a UTC package could be assembled on predefined footprints, and wire interconnects are printed during the fabrication of large-area components.

1.2.3 Flexible Substrate Materials

Electrical, mechanical, and thermal material properties need to be carefully considered while choosing the substrate material for flexible systems, as no optimum material is currently dominating. Polyimides (PI) and PDMS are often used, depending on the required thickness and stretchability, and metallic foils are applied to solar cell manufacturing. The structural and dimensional stability of the substrate material during different manufacturing processes is crucial for processing uniformity and improving yield. The substrate flexibility should be maintained at different operating conditions, as some polymers tend to become brittle at low temperatures. Having low moisture absorption can contribute to material stability and avoids sudden delamination.

Thermal resistance is important from the manufacturers' perspective because it sets the upper limit on the processing and assembly temperature. It is also important for the customer to know the temperature range, especially in temperature-sensing applications in which the sensor linearity and response time are linked to the packaging and substrate properties. Substrate materials with high tear resistance are less vulnerable to hardware hacking that infiltrates data privacy. Dielectric properties such as low dielectric constant and dissipation factor are needed for frequencies beyond 100 MHz. As the final thickness of flexible electronic systems becomes thinner, much focus should be directed to compatible adhesive materials with stable electrical, mechanical, and thermal characteristics.

1.3 Active and Passive Components for HySiF Integration

Electronic components that are incorporated into HySiF have structural specification restricted by the targeted application. For instance, conventional surface mount passives are not readily compatible with the HySiF concept. Therefore, HySiF components are designed to minimize external components and to be self-contained (i.e., during their operation, they do not depend on external auxiliary components, such as decoupling capacitors, matching resistors, or clamp diodes). As an example, certain wireline communication protocols, such as I2 C, require pull-up or pull-down resistors for their proper operation. For simplifying the HySiF integration, such resistors should be moved from on-foil to be on-chip. Alternatively, if this is needed, thin-film passives can be directly fabricated on the flexible substrate, which then leads to another process step in addition to having wide variations in the value of the passives since lithography techniques with more tolerance are used.[26], [27]

In the following sections, several active and passive electronic components that are readily available for HySiF integration are briefly discussed.

1.3.1 Ultrathin Chips

Since the active part of conventional CMOS ICs takes about 1% of the total thickness, chip thickness is a degree of freedom independent of the conventional CMOS scaling. Within the HySiF concept, UTCs can be used as a high-performance centralized SoC. Complex tasks, such as sensor readout, high-speed DSP, wireline and wireless communication, power management, and machine learning on the edge, can be packed inside CMOS UTCs. In this way, complex electronic tasks are assigned to the well-established CMOS technology, and the number of embedded silicon UTCs is minimized per HySiF.

1.3.2 Flexible Sensors

Large-area off-chip/on-foil sensors or sensor arrays can be distributed on the foil, as they complement the operation of the miniaturized on-chip electronics. Several boundary conditions regarding sensor protection, processing conditions, low-delay interconnects, reliable operation, low hysteresis, and smooth versus rough substrate surface are considered while developing sensors for HySiF integration and thus realizing smart flexible sensor systems. For instance, the electrical properties (i.e., electron/hole, ionic, or proton conductivity) of resistive thin-film sensors are influenced by the activity and properties of the underling flexible substrate. This effect can be mitigated by employing bridge configurations, special layout techniques for stress compensation.[28], [29] Using capacitive sensors instead (especially parallel-plate capacitive sensors) is highly recommended for HySiF integration, as inherent stress insensitivity is possible.

Multimodal sensors (i.e., single sensor elements that are able to sense different physical parameters at once) are on the rise. They can be utilized within the HySiF concept to simplify the fabrication process steps and optimize the surface area. For instance, a ferroelectric gate insulator is integrated into the gate oxide of an organic TFT.[30] Ferroelectricity includes both pyroelectricity and piezoelectricity, where if we establish a tensor relation for the TFT electrical parameters, the temperature and mechanical strain can be simultaneously detected. Besides, 3D integration of thin-film sensors by vertical stacking is another possibility of achieving multimodel sensors.[31], [32]

Other bendable sensors are battery-free and can operate and transmit their sensing data wirelessly.[33] They can passively use an energy field (electromagnetic field of NFC-enabled device) to encode their sensing data by modulating this field. For instance, an ultrathin, highly sensitive (gauge factor > 14,000) $Ti_3C_2T_x$ MXene resistive strain sensor is coupled to an NFC-enabled device.[34] Such unconventional sensing materials can significantly enhance the performance of electronic skins, soft robots in particular.

1.3.3 Metal-Oxide and Organic Thin-Film Electronics

A new paradigm in flexible electronics is enabled by the field of thin-film electronics, as it offers, among other things, inherently thin and bendable devices and a simple room-temperature and quick fabrication process. In addition, organic thin-film electronics are potentially more environmentally friendly compared to the silicon technology. Within the HySiF concept, thin-film electronics, including metal-oxide and organic electronics, complement and operate alongside the silicon UTCs. Due to their amorphous nature, they can be fabricated on a wide range of substrates. However, their material properties and semiconductor electrical properties are inevitably impacted by the surface quality of the flexible substrate.

Using thin-film transistors (TFTs), sensor array addressing and signal multiplexing are no longer limited by the I/O count of the silicon UTCs. For instance, a large-area array of frequency-hopping ZnO TFT oscillators has been implemented in [6] in order to extend the number of available sensor channels from the typical quadratic behavior (N^2) of active matrix addressing to an exponential behavior (2^N).

In the field of neuroscience, particularly for in vivo intracellular and extracellular recordings, OLEDs and TFTs have been integrated into the tip of multielectrode arrays for photo stimulation and signal amplification, respectively.[35], [36], [37] Here, TFTs offer high interface capacitance, which results in high SNR.[37] Furthermore, two-dimensional electronics (e.g., using single-layer MoS_2) can be utilized to achieve large-scale flexible electronics systems. [38]

1.3.4 Flexible Antennas

Several flexible antennas have been reported, and their performance is steadily approaching that of their rigid counterparts.[39], [40] The electrical behavior of antennas, flexible antennas in particular, is strongly coupled to the electromagnetic properties of the substrate, as well as the passivation superstrate. Low electrical permittivity and dielectric loss in the vicinity of the antenna tend to improve its efficiency. Further discussion of specific demonstrators and properties of the antenna packaging is presented in Sections 2 and 3.

1.3.5 Flexible Displays

For ultrathin flexible display, the market is currently expanding as foldable portable electronics, notably smartphones, are on the rise. The Chinese company Royole is currently mass-producing AMOLED flexible displays, which are lightweight, 100 μm thick, and can be bent down to 1 mm radius. However, the economic feasibility of integrating flexible displays into HySiF, such as smart labels or e-skins, is still in question.

1.3.6 Energy Harvesting and Storage

For autonomous and self-powered HySiF, local energy storage is needed. On the bright side, energy-harvesting devices, such as photovoltaics and electromagnetic field harvesters,[41], [42] can be rendered ultrathin and bendable. On the other side, the form factor of energy storage devices is considered a bottleneck for the development of a portable seamless HySiF.

Flexible lithium-ion batteries have been reported.[43], [44] However, size-able form factor, low power density, and heat generation are the main concerns. Complementary storage devices, such as flexible supercapacitors,[45], [46] can be used to supply high power demands but for limited duration.

1.4 System-Level Concept

Figure 1.3c shows a block diagram for a generic model reflecting the envisioned HySiF concept in which main electronic subsystems are defined. By considering a specific application, this block diagram can be customized (number of SoC UTCs or removal of flexible display for economic implementation). In this HySiF concept, a centralized SoC UTC performs accurate sensor readout, digital signal processing, communication, and power management. In this way, the number of embedded silicon UTCs is limited to one to three ICs per HySiF.

Moreover, large-area, off-chip sensor arrays are distributed on the foil. Here, on-foil sensors require signal conditioning circuits, which eventually brings the signal flow back to the centralized SoC. Optionally, thin-film transistors can be employed to preamplify and multiplex the sensor signals. Therefore, signal integrity is preserved, and the number of on-foil sensors can be extended beyond the limits of the SoC I/O pins.

Wireless microcontroller UTCs are good candidates for combining high-speed computational power with multi-standard wireless connectivity. For ultralow power applications, bespoke microcontrollers on plastic substrates are good alternatives, since their implementation is optimized for the target application.[47]

2 Ultrathin Chip Fabrication and Flexible Packaging Technologies

For fitting more functionality in a given volume (or a given set of functions in a reduced volume), technology scaling is traditionally the hallmark of the semiconductor industry. However, since the active part of the chip usually takes only about 1% of its total thickness, chip thickness is a degree of freedom independent of conventional scaling. In this section, current applications of ultrathin chips (UTCs) and manufacturing techniques targeting less than 50 μm chip thickness are discussed. Additionally, recent chip packaging techniques are reviewed with a special focus on UTC flexible packaging.

2.1 Ultrathin Chip Process Technologies

A chiplet is a tiny chip that integrates generally reusable intellectual property (IP) blocks, which can be combined with other chiplets in an SiP to form a new module or platform. A shift from the traditional monolithic SoCs to chiplets-based MCMs and SiPs results in a cost-efficient solution, provided that reliable, high-speed inter-chiplet wire interconnects are used.

As an example, Xilinx introduced Versal ACAP (Adaptive Compute Acceleration Platform) in which 3D stacked chiplets form a heterogeneous integration of three programmable engines.[48] Intel Foveros is also a 3D integrated combination of CPUs, GPUs, and application processor chiplets. [49] ODSA (Open Domain-Specific Architecture) is an open-source initiative, which aims at enabling lower-cost customizable chiplet-based solutions competing against conventional SoCs.[50]

With that in mind, UTC manufacturing is considered a cornerstone for achieving this new level of SiP integration. Traditionally, UTC fabrication is divided into subtractive and additive or post-processing and preprocessing techniques.[51] Without loss of generality, subtractive techniques seem simpler for fabricating UTCs, though they cause a waste of materials that are difficult to recover. On the contrary, additive techniques optimize material usage to achieve economic UTC fabrication. However, they are also considered a risk for chip foundries, as preprocessed wafers might deviate from standard wafer properties, which poses unexpected challenges that could potentially disturb uniform manufacturing throughput and yield. In general, additive UTC technology leads to lower cost-of-ownership (CoO) toward smaller die thickness, whereas CoO increases for subtractive techniques.

2.1.1 Subtractive Approaches

UTCs with precise thickness can be achieved using silicon-on-insulator (SOI) wafers, as the buried oxide acts as an etch stop layer during the selective etching of the silicon wafer backside. However, this sets certain boundaries on the technology choice, development costs, and other challenges related to thin wafer/chip handling. Alternative dry etching techniques for bulk silicon wafers use reactive ion etching that targets the wafer backside with UTC thickness of about 20 μm.[52], [53], [54]

Wafer back-grinding is another conventional technique to fabricate thin chips targeting applications, such as RFID-based biometric passports from the company NXP with typical thicknesses of less than 200 μm.[55] However, using traditional grinding to achieve UTCs is limited by degradation in the structural integrity of the wafer, as crystalline defects and micro-cracks start to propagate. Figure 2.1a illustrates the dicing-before-grinding (DBG) method,[56], [57] developed to circumvent the challenges associated with conventional grinding. UTCs with thickness down to 20 μm and improved chip strength are achieved using the DBG process.[56], [57]

However, thin wafers are prone to mechanical damage, thus potentially lowering yield dramatically. As a consequence, the TAIKO partial back-grinding method was developed.[58] Here, the back-grinding is applied only to the inner part of the wafer backside, leaving a thick and stable ring-shaped frame. Other wafer subtractive techniques include layer transfer of the active side of the wafer, such as in proton-induced exfoliation,[59], [60] and controlled spalling technology achieving a nanoscale substrate thickness of 60 Å.[61]

2.1.2 Additive Approach

Additive techniques for manufacturing UTCs require preprocessing steps for wafer epitaxial preparation by mainly introducing one or several porous silicon layer(s). ChipFilm technology is a known example of fabricating 10 μm-thick UTC with well-controlled uniform thickness and minimum defects. Figure 2.1b illustrates the main pre- and post-processing steps for UTC fabrication using ChipFilm technology.[62] A by-product of achieving ultrathin form factor is an increase in the freestanding chip warpage. ChipFilm dies exhibit slightly lower warpage compared to the DBG method, which is affected by stress induced with the grinding. [57] The warpage can be tailored using suitable layout density design rules and detailed layer stress management for the CMOS manufacturing technology.

2.2 Flexible Electronic Components Packaging

Assembling rigid passive and active electronic components on or in flexible substrates has been widely used in packaging and PCB technologies in what is known as Flex-PCBs. A review of packaging technologies targeting inherently flexible electronic components, such as UTCs, flexible sensors, and antennas, is presented in this section.

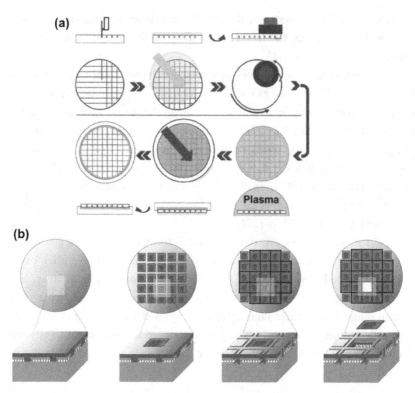

Figure 2.1 (a) Wafer back-grinding using DBG process.[56] From left to right, half-cut dicing, lamination of surface protection tape, back-grinding down to engraved cuts level, plasma stress-relief etching, standard dicing frame mounting, protective tape removal, and chip assembly by the pick-and-place process. (b) Schematic of ChipFilm technology main processing steps.[62] A porous layer is etched, followed by formation of coarse pores from which a hall of pillars is formed. Next, standard CMOS integration is performed. Afterwards, etch trenching reaches the underlying cavity, leaving only the pillars holding the chips. Finally, the pick-and-place tool snags the chip from above.

The main purpose of flexible packaging is to mechanically support ultrathin electronic components by their assembly at the least-stress location (as long as it is not a device that needs elevated stress levels, such as a stress sensor or a strain gauge), which is usually the neutral plane of stress residing in the middle of a uniform packaging substrate. In a composite substrate, that neutral line of stress will be substantially away from the middle of the substrate when the composite materials feature significantly different E-moduli.

2.2.1 Flexible Sensors Packaging

Whether it is an environmental, motion, wearable, or chemical sensor, the protection, interconnection, and passivation of the sensor without impacting its performance is a cumbersome optimization task for the flexible package. Conventionally, the sensor surface is protected with a superstrate material similar to that of the flexible substrate. Figure 2.2a shows a typical structure of a flexible package for sensor applications. Material properties such as thermal resistance, moisture uptake, and elasticity need to be considered in the choice of a sensor packaging material.

For electrical contact, vias through the superstrate are usually etched and noble metal pads are fabricated, which usually feature relaxed size and pitch, as the typical pin count per sensor is about 2–4 pins. As an example, a 28 µm-thick ultrasonic transducer array sheet using piezoelectric polyvinylidene fluoride (PVDF) film is covered with NiCu metal film as a common ground and topmost material.[63] A flexible absolute pressure sensor is fabricated on a PI substrate and is sealed using an Al_2O_3 layer that is covered with PI as a superstrate.[64]

2.2.2 Flexible Antenna Packaging

It is not straightforward to use metallic foil for packaging flexible systems that include wireless connectivity, as electromagnetic interaction with the package body increases and, thus, distortion of the transceiver's high-frequency signals

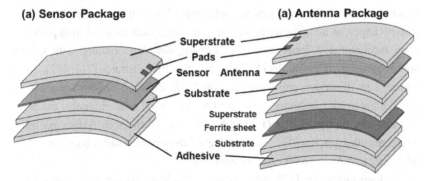

Figure 2.2 Illustration of flexible packaging solution for (a) sensors and (b) antennas. Usually, sensors and antennas are placed in the neutral stress plane inside a polymeric substrate and chip package using spin-on superstrate composite. Contact pads are most often fabricated on the superstrate using solderable metals such as Cu or noble metals such as Au. An additional stack of ferrite sheet is placed below the antenna, as shown in (b), for better RF signal reception.

occurs. Insulators are therefore preferred as packaging material, since lower electrical permittivity (or dielectric constant) and dielectric loss tangent in the vicinity of the antenna tend to improve its efficiency. PI and PDMS are widely used substrate materials with dielectric constants and loss tangents in the range of (ε_r / δ) 3.2–3.7 / 0.0018–0.0026 and 2.3–2.8 / 0.0015–0.0030, respectively. Figure 2.2b illustrates a typical structure of a flexible package for NFC antennas.

SNR degradation is faced when attaching the flexible antenna to metallic objects or batteries. A ferrite sheet or a combination of flexible metamaterials and ferrite films could be placed under the flexible antenna.[65], [66] Printed antennas have been fabricated on flexible substrates with a near-constant performance during bending; however, their surface is not passivated.[67], [68] Antennas for 915 MHz and 2.5 GHz transmission have been fabricated by embedding conductive fibers in PDMS substrates, thus demonstrating the possibility of realizing bendable and stretchable antennas for flexible electronics applications.[69], [70] A challenge remains concerning the stability of the antenna properties and tuning to the mechanical activity of the targeted application.

2.3 Ultrathin Chip Flexible Packaging

Conventional chip packaging relies on wire-bonding and flip-chip assembly with solder bumps technologies for realizing the electrical connection between the on-chip fine-pitch I/O pads and the coarser pins of the package. However, typical wire-bonding techniques are not compatible with mechanically flexible chip packages, as bonding wires will be dismantled from either chip or package upon mechanical deformation. Alternatively, *flip-chip bonding* with solder bumps is used to assemble UTCs in flexible PI substrates.[71], [72], [73] Figure 2.3a illustrates one variant of the UTC flip-chip bonding process in which anisotropic conductive adhesive (ACA) is used to interconnect the 50 × 120 μm² Au bumps on the UTC to Ni/Au pads on the foil. Achieving finer interconnection pitch will require better chip alignment, smaller bump area and thickness, and ACAs with high conductivity.

The long history of PCB manufacturing has also inspired UTC packaging solutions such as *embedded component technology (ECT)*, originally developed for 3D integration and embedding chips and passives in rigid PCBs. UTCs are assembled either face-up or face-down on PI or liquid crystal polymer (LCP) substrates, and micro vias ($\approx$ 100 μm) are created using laser drilling.[81], [82], [74] To avoid chip damage, electroplating of a Ni/Pd protective layer on the chip pad is used.[74]

Advances in metal *printing technologies* enable simple and fast UTC packaging techniques that can be extended in the future to larger areas and mass production. UTCs are assembled on flexible substrates where narrow tracks are printed by inkjet [31] or aerosol jet,[30] using silver inks. Silver inks provide higher conductivity but are susceptible to cracks; conductive polymeric inks, however, are more stretchable but have much lower conductivities. In the case of face-up chip assembly, dielectric leveling ramps might be needed to provide a gradual transition from the UTC surface to the substrate fan-out pads (see Figure 2.3b).[74]

2.3.1 ChipFilm Patch (CFP) Technology

The previously discussed UTC flexible packaging techniques exploit commercially available polymeric and metallic sheets or foils such that UTCs are mostly assembled between successive layers of the same material. However, spin-on polymers, which are commercially available in liquid form, provide more control on the package composition and the thickness of each layer. As shown in Figure 2.3c, the *ultrathin chip package (UTCP)* is realized by embedding UTCs in spin-on PI layers with a thin polymer layer (e.g., benzocyclobutene [BCB]) for chip adhesion.[76]–[80] A wireless electrocardiogram (ECG) sensor system uses UTCPs (sensor readout, microcontroller, and RF communication ICs) as interposers, which are embedded in a Flex-PCB, demonstrating the feasibility and compactness of this flexible packaging and integration technique.[83]

Another CMOS-compatible UTC embedding technology, known as *ChipFilm Patch (CFP)*, allows for < 10 μm fine-pitch fan-out interconnection.[84] CFP is

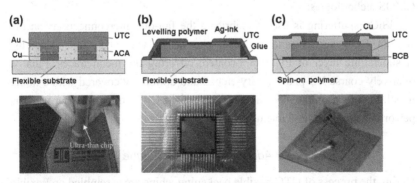

Figure 2.3 Cross section and photographs of UTC flexible packaging technologies based on (a) flip-chip bonding using anisotropic conductive adhesive (ACA),[71], [72], [73] (b) inkjet or aerosol jet printing,[74], [75] and (c) ultrathin polymers.[76], [77], [78], [79], [80]

based on conventional IC manufacturing processing technologies (e.g., AlSiCu metal sputtering) in combination with layers of spin-on polymers (e.g., spin-coating of PI/BCB stacks) to create one or more interconnect layers. The fine-pitch interconnectivity of CFP allows for implementing pad-limited, high-performance ICs into flexible electronic systems. Similar to UTCPs, CFP can be used as an interposer.[85] Furthermore, CFP can be used as a stand-alone system in which various active and passive flexible electronic components are manufactured on or into the CFP foil.[86]

Figure 2.4 shows the cross-section schematics of the CFP processing steps in which two polymers are used for embedding UTCs.[87] Following the CFP concept, a temporary adhesive and stress-relief layer are fabricated on a silicon carrier wafer.[88] A reinforcement substrate is fabricated as PI is spin-coated directly on the previous layer. A BCB layer is then spin-coated in which a cavity resembling the UTC is structured and etched. An additional thin BCB layer is spin-coated, on which the UTC is assembled in face-up orientation. Note that another CFP variant adopting the face-down approach is discussed in [19]. After the BCB curing process, another BCB layer is spin-coated to completely encapsulate the UTC. Dry etching using BCB or AlSiCu as a hard mask is used to create micro vias on top of the UTC I/O pads.

The optimization of the etching process is crucial for the CFP process. A mixture of CF_4 and O_2 gases is used during the plasma etching of the BCB mask. Micro vias having 10–20 μm depth are etched in the CFP process. Chip warpage, nonuniform BCB/PI topography, process-induced stresses, and air bubbles are nonidealities, which are directly linked to the quality and shape of the micro vias opening and sidewall slope. Optimized standard via sizes are then used throughout the CFP process, similar to standardized micro vias in modern CMOS technologies.

AlSiCu sputtering is used to fabricate the fan-out interconnections on the smooth BCB surface and to cover the micro vias and their sidewalls. A superstrate layer of PI is spin-coated, covering the metal traces. Finally, relatively coarse I/O pads are dry-etched and optionally covered with a noble metal. Finally, a simple mechanical release of the CFP from the carrier wafer is performed using either a manual or a laser cutter.

2.3.2 Adaptive Layout Technique

During the process of UTC flexible packaging, chips are assembled on flexible substrates using component placers or die bonders. Inaccuracies during chip placement, such as positional or rotational offsets, impact the overall accuracy of the chip embedding and interconnection technology. In addition, during CFP

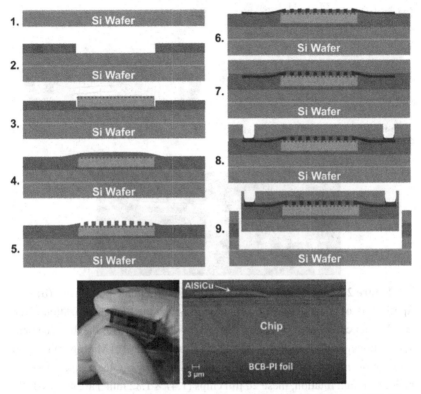

Figure 2.4 Illustration of ChipFilm Patch (CFP) fabrication process.[87] (1) Adhesion lowering layer and wafer-bow reduction on silicon carrier wafer, (2) spin-coating of PI and BCB, (3) chip assembly inside a prestructured cavity resembling UTC size, (4) chip embedding with BCB, (5) structuring top BCB layer for via opening, (6) AlSiCu metallization and structuring, (7) covering with BCB and PI superstrate, (8) outer pad opening and deposition of noble metal layer, and (9) releasing CFP from the carrier wafer. Image of ChipFilm Patch package with embedded UTC and SEM micrograph of the CFP cross-section showing, from bottom to top, the following layers: BCB/PI flexible substrate, 20 μm chip, chip pad, thin BCB, 1 μm AlSiCu metallization, and BCB/PI thick passivation layer.[87]

manufacturing, UTCs tend to move within the predefined cavity during the curing of the spin-on polymers. Figure 2.5a illustrates inaccuracies in chip placement during HySiF and CFP manufacturing.

Adaptive layout technique is applied to the CFP flexible packaging technology in order to extend its fine pitch-interconnection capabilities.[89] The coordinates of each embedded UTC are measured by utilizing on-chip alignment marks and a maskless laser direct writer, which is normally used for lithography steps in CFP manufacturing; the overlay accuracy is 200–500 nm.

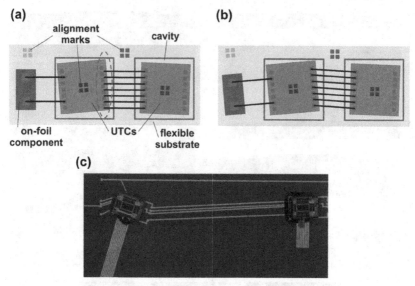

Figure 2.5 (a) Illustration of inaccuracies during chip placement. (b) Application of adaptive layout technique in which the first metallization layer and all other consequent layers are adjusted to compensate for the displacement and rotational errors originating during UTC embedding. (c) Micrograph of two logic chips, integrated into CFP using the adaptive layout technique. Despite intentional high rotation, these 20 μm chips (1.47×1.82 mm^2) are successfully interconnected.[89]

Layouts for the first metallization and all consequent layers are updated in order to preserve the original circuit interconnects, as shown in Figures 2.5b and 2.5c. The alignment of on-foil components, such as sensors, TFT circuitry, and antennas, is either adjusted to the orientation of the associated chip or is left unadjusted, provided that the adjustments in the layout of the respective wire interconnects are sufficient. Note that the overall accuracy of the CFP is now linked to the structuring as well as to the overlay accuracy of the lithography tool.

Chip warpage, nonuniform BCB/PI topography, and process-induced stresses are nonidealities that limit the accuracy of the CFP process and the adaptive layout technique. Therefore, multiple on-chip alignment marks are then needed to ensure reliable misalignment measurements. Typical chip warpage of a stand-alone 20 μm UTC is about 30 μm, which matches with the upper limit of the focusing capabilities of the utilized laser writer (about 40 μm).[89] Note that the chip warpage for stand-alone and partially embedded 20 μm UTCs diminishes gradually from 30 μm to less than 10 μm during CFP processing.

3 On-Foil Passive and Active Components

Currently, most flexible electronic components are either fabricated as stand-alones or combined with bulky sensor readout and/or wireless communication silicon chips glue-attached to the surface of a flexible substrate.[6], [7], [8] Other recent approaches use only one technology to achieve a truely bendable sensor system (e.g., an integrated temperature sensor and A/D converter using amorphous Indium-Gallium-Zinc-Oxide [InGZnO] thin-film transistors).[90] However, HySiF targets an economic implementation of sensor systems by ensuring compatibility and combining the mature CMOS chips with the large-area passive and active components.

Table 3.1 establishes a comparison between very-large-scale integrated on-chip sensors and their large-area on-foil counterparts. It is obvious that an array or matrix formation is more easily realized when sensors are printed in contrast to having to deal with multiple silicon chips. However, signal integrity is

Table 3.1 Comparison between on-chip and on-foil sensors concepts.

	On-chip sensors	On-foil sensors
Array formation	complex	simple
Interconnect parasitics	low	high
Readout electronics	integrated	usually remote
Accuracy	high	low to medium
Mechanical flexibility	only with thinned chips	highly flexible
Form factor	small	small
Surface area	small	larger
Thickness	thicker	thin
Thermal contact resistance	high when in package	low
Lithography	complex	simple
Fabrication temperature	high	lower
Power consumption	low	higher
Reliable lifetime	years	months
Biodegradable	no	possible
Substrate	silicon	polyimides, Kapton, paper, ...
Package	plastic, ceramic, foils, ...	polyimides, Kapton, paper, ...

negatively influenced when long metal interconnects are used between large-area sensors and a central interface and processing chip.

In this section, on-foil active and passive components, which are suitable for HySiF integration, are discussed. As a benchmark, environmental sensors such as temperature and humidity sensors, printed strain gauges, on-foil inductor loops for NFC communication, and simple on-foil digital circuits using organic thin-film transistors (OTFT) will be presented.

3.1 Ultrathin Environmental Sensors

Environmental parameter monitoring is a basic requirement for most IoT application scenarios. Whether it is a sensor node in a smart factory, or a health-monitoring wearable device, or an RFID tag for logistics sorting, temperature, humidity, and pressure are crucial physical parameters that need to be closely monitored. A sudden change in the measured readings certainly indicates an event that is worth recording and investigating. Commercial environmental sensors are highly integrated systems-on-chip that are usually available in plastic or ceramic packages. Although their form factor is in the mm range, it is rather inconvenient to integrate them in applications where full mechanical flexibility and large area coverage are needed. Recent flexible temperature and humidity sensors have achieved a relatively low detection limit and high degree of stability.[91], [92] However, the readout electronics and signal processing are almost exclusively done remotely and by means of a different substrate.

3.1.1 On-Foil Temperature Sensor

For accurate temperature sensing, a low and linear thermal contact should be maintained between the sensor and the surface to be measured (thermal interface A in Figure 3.1). Although on-chip temperature sensors are highly accurate,[93] the thermal insulation of the CFP package hinders a good thermal contact between the embedded chip and the surface to be measured. However, on-chip sensors can be used to monitor the chip self-heating and in-package thermal behavior (thermal interface B in Figure 3.1). In order to resolve this issue, on-foil thin-film metal temperature sensors are used.[94], [23]

Sensor Design and Fabrication

Figure 3.1 shows the cross section of the chip-embedding process using CFP together with an on-foil temperature sensor. Following the CFP process,[89] the flexible substrate is composed of a stack of benzocyclobutene (BCB) and polyimide (PI) layers. Both polymers are spin-coated onto a silicon carrier wafer. Several spin-coating/baking steps are performed to achieve the required

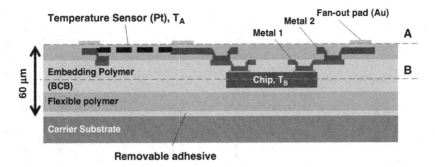

Figure 3.1 Cross section of one variant of the CFP process where ultrathin chip and on-foil temperature sensors are integrated on a flexible substrate.[95]

substrate thickness. After the final coating step, thin-film platinum (Pt) is used to fabricate a resistance temperature detector (RTD) on the flexible substrate surface. The thin form factor and small thermal mass of the on-foil thin-film Pt wire allow for effective thermal management.

Test sensor wafers are fabricated for the electrothermal sensor characterization. Figures 3.2b and 3.2c illustrate the sensor design, where two different layout techniques are applied, using serpentine and horseshoe structures. The former is used when a compact sensor surface area is needed, while the latter is applied when stress compensation is crucial. Here, the resolution is limited by the standard laser direct writing. Figures 3.3f and 3.3g show the RTDs when used as unit elements in a Wheatstone bridge configuration. The purpose of this experiment is to assess the impact of the fabrication process deviations on the resistance mismatch. The bridge offset resistance value R_{offset} is calculated as follows:

$$R_{offset} = \frac{V_1 - V_2}{I_{Bias}},$$

where V_1 and V_2 are the voltage difference between the outputs of the bridge and I_{Bias} is the bridge biasing current. The horseshoe-based Wheatstone bridges show superior matching when compared to the serpentine-based bridges. This is attributed to better randomization of the fabrication process variations for the horseshoe layout, whereas, in the fabrication of the serpentine structure, it is mostly done along one axis.

Sensor Characterization

Figure 3.2a shows three different fabrication setups that are used to characterize the on-foil temperature sensors.[95] In the first setup, a 50 nm/10 nm platinum/ titanium (Pt/Ti) sensor stack is fabricated using the liftoff process on a 680

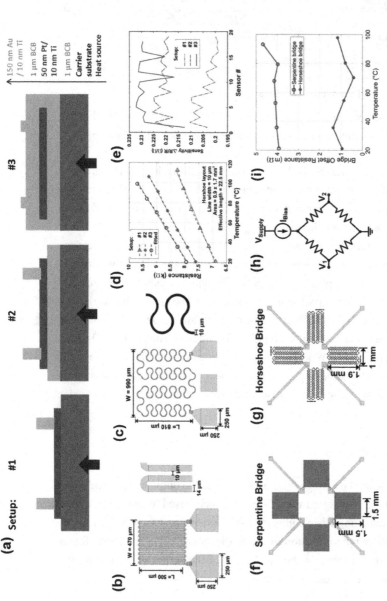

Figure 3.2 (a) Cross section of three setups used to electrically and thermally characterize the thin-film Pt temperature sensor.[23] (b) Serpentine and (c) horseshoe sensor layouts are used, respectively. The line width and spacing of sensor structure are limited by the lithography technique. Wide pads are used to allow for an accurate four-point measurement of the sensor response. (d) Measured resistance of three temperature sensors with similar dimensions from three different wafers. (e) Measured sensitivity of multiple sensors from the three different setups shown in (a). Wheatstone bridge configuration using (f) serpentine and (g) horseshoe RTD layouts. Using the measurement setup in (h), the measured offset resistance is plotted against the temperature in (i). The horseshoe layout shows more than four times less offset due to a better randomization of the fabrication process non-idealities.

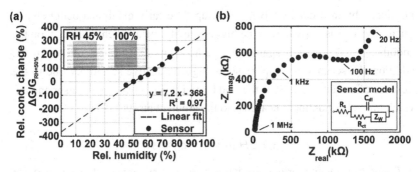

Figure 3.3 (a) Measurement results of the electrochemical sensor relative conductance change with relative humidity.[99] A linear fit is also plotted with a coefficient of determination (R^2) = 0.9724. The inset illustrates the change in the film optical thickness (surface color) at RH of 45% and 100%. (b) Nyquist plot of the electrochemical sensor at RH of about 60%. The measurement results indicate that the sensor electrical behavior can be modeled using Randles equivalent circuit,[8] which is shown in the figure inset. The model consists of the electrolyte resistance (Rs), charge transfer resistance (Rct), specific electrochemical element of diffusion (Zw), and double-layer capacitance (Cdl).

μm-thick SiO_2/Si wafer. For the second and third setups, the same sensor stack is fabricated on a 1 μm-thick BCB layer, while using Si wafers as a carrier. In addition, for the third setup, the sensor stack is covered with a layer of 1 μm-thick BCB. A thermo-chuck is used as a heat source, and its temperature is varied, starting from 40°C to 120°C in 20°C steps. At room temperature, an extra data point is measured. The topmost surface temperature (in setup #1, SiO_2; in #2 and #3, BCB) is monitored using a reference Pt100 foil sensor. The maximum inaccuracy in the quasi–steady-state temperature measurement is ±0.75°C. Note that increasing the thickness of the CFP embedding stack improves the thermal insulation between the on-foil sensors and any heat source embedded inside the flexible package.

Different RTD surface areas are measured ranging from 0.06 mm² (80 Ω) to 2.5 mm² (50 kΩ). Figures 3.2d and 3.2e show the static characterization of three RTDs with similar dimensions but from different wafers, following the experiments previously described in Figure 3.2a. The four-point measurement technique is used to connect the RTD under test with precision source measurement units (SMUs). Note, that the base resistance value variation depends on the substrate material and its surface topography. The extracted sensitivities ($\Delta R/R_{20°C}$) of 18 different sensors are plotted in Figure 3.2e. Although the surface temperature in setup #2 is less than that in setup #1, due to the heat loss introduced by the added BCB layer, setup #2 clearly exhibits a better sensor sensitivity. This is attributed to the higher-quality metal when fabricated on the smoother BCB surface against the silicon oxide surface of setup #1. Additionally, sensors in setup #3 show higher effective sensitivity due to the presence of an additional BCB thermal barrier layer between the sensor and air, which limits the nonlinear heat loss to the ambient air.[96].[97]

3.1.2 Humidity Sensors

Different transducing principles have been investigated and developed for relative humidity sensing. Capacitive humidity sensors have dominated the market and are widely used, since no static power consumption occurs during operation. However, using conductive polymers (CP), graphene oxide, carbon nanotubes, and nanowires as humidity-sensing materials has recently gained interest at many research institutions, due to their flexible tuning and very high sensitivity. Table 3.2 reviews the commonly used humidity-sensing materials and their properties. In the following section, two on-foil humidity sensors, namely an electrochemical and a conventional capacitive humidity sensor, are investigated.

Electrochemical Humidity Sensor

Most commercially available humidity sensors have a relatively slow response time (> 5 seconds), low accuracy (< ±3% RH), and low sensitivity that requires a higher-precision interface circuit. Humidity sensors with both huge sensitivity and fast response time are needed not only as environmental humidity sensors but also as pixels in touchless screens and positioning systems. In [98], nanosheets composed of phosphatoantimonic acid ($H_3Sb_3P_2O_{14}$) were spin-coated on rigid quartz, which achieved a proton conductance change of about five orders of magnitude ($10^{-10} – 10^{-5}$ S, 0–100% RH) and response time of less than two seconds.

Sensor Fabrication

The same nanosheet material is spin-coated on a BCB/PI flexible substrate. The nanosheets are spin-coated at 2,000 rpm for 90 seconds, reaching a thickness of about 200 nm.[99] AlSiCu electrodes were already sputtered on the BCB/PI substrate, defining a sensing area of 0.5×4 mm^2. Electrical as well as optical readouts are possible. As water molecules get intercalated between the nanosheets, variation in the film thickness causes a twofold effect: a change in the proton conductance, and a shift in the film refractive index.

Table 3.2 Review of recent relative humidity sensing techniques and materials.

Ref.	Transduction	Sensing Material	Fabrication	Sensitivity	Range (%)	Response time (s)	Thickness	Substrate	Electrode material
[99]	**C**	**PI**	**Spin-coating**	**±90%**	**60–80**	**6**	**3 μm**	**PI**	**Au**
[99]	**EIS / R**	**$H_3Sb_3P_2O_{14}$**	**Spin-coating**	**±368%**	**45–80**	**2**	**200 nm**	**PI**	**Au**
[157]	C	PI	Spin-coating	15.2fF/%RH	5–85	80	4.6 μm	Glass	Au/Cu
[158]	C	Kapton	Inkjet printing	0.5F/%RH.mm^2	20–90	350	75 μm	Kapton	Ag-ink
[159]	C	PEG/PEDOT:PSS	Spin-coating	0.0085%RH	10–90	2	250 μm	PDMS	PEDOT:PSS
[160]	R	PANI/PEDOT:PSS	Inkjet printing	200%	16–98	420	100 μm*	Paper	PEDOT:PSS
[161]	EIS / R	GO	Spray-coating	4.7MΩ–0.6kΩ	25–95	0.5	12 μm	Si/SiO2	Au/Ti
[162]	R	GO	Drop casting	1200x	25–88	5	2.5 nm	Si/SiO2	Au
[163]	C	GO	Drop casting	37800%	15–95	10.5	-	Si/SiO2	Au
[164]	R	WS$_2$	W + Sulfurization	2357x	20–90	5	2.1 nm + 55 μm	PDMS	Graphene
[165]	EIS / R	WO$_3$	Spin-coating	276.8%	11–95	6	171 μm + 500 μm	Ceramic	Ag-Pd
[166]	R	BP/Graphene	Electro-Spraying	43.4%	15–70	9	-	Si/SiO2	Au
[167]	R	nanofibres	Drop casting	45000x	5–80	2.2	20 nm	Glass	Au
[168]	EIS/C	IESM	Natural	5–0.5MΩ/32–60pF	0–90	2	19 μm	IESM	Ag-ink

Sensor Characterization

The sensor is characterized in a climate-controlled chamber in the range of 45–80% RH at a constant temperature of 30°C. Figure 3.3 shows the measured static response and Nyquist plots of this electrochemical humidity sensor.[99] In the measurement range of 45–80% RH, the electrochemical sensor relative conductance change ($\Delta G/G_{RH=50\%}$) is about 370%. This is measured using an LCR meter with an input sinusoidal signal of an amplitude and frequency of 300 mVpp and 800 Hz. By extrapolating the measured sensitivity to the full humidity range, a slight degradation in the nanosheets' quality is observed when fabricated on the BCB/PI substrate compared to the rigid and smoother quartz. This indicates that BCB/PI is a suitable substrate for HySiF applications. Response time of less than two seconds is measured when the sensor is repeatedly saturated to 100% RH. Note that such an electrochemical sensor requires a complex readout technique based on electrochemical impedance spectroscopy, which might hinder the sensor's practical use in low-power sensor systems.

Capacitive Humidity Sensor

Capacitive humidity sensors have recently dominated the market due to the robustness of the humidity-sensitive polymers used as dielectric material in addition to the fact that no static power is needed to operate the sensor.

Sensor Fabrication

Here, using the same BCB/PI substrate, a capacitive humidity sensor is fabricated using Durimide PI as dielectric material spin-coated on interdigitated electrodes.[99] The AlSiCu electrode finger width and spacing are 10 μm and 20 μm (see Figure 3.4d), respectively. The PI thickness is about 2.7 μm, achieved using a spin-coating speed of 1500 rpm and a curing temperature of 250°C for 120 minutes. As shown in Figure 3.4a, a reference capacitor with the same design is placed in close proximity. A thin layer of BCB is spin-coated and structured only on top of the reference capacitor in order to shield it from any moisture uptake. This reference capacitor together with the sensor capacitor are used to build a half-bridge configuration in a flexible smart sensor, as explained in Section 4.4.

Sensor Characterization

Figure 3.4b shows the fabricated on-foil humidity sensors on the silicon carrier wafer. For sensor characterization, the structures are measured once on the carrier wafer and another time after their mechanical release from the carrier.

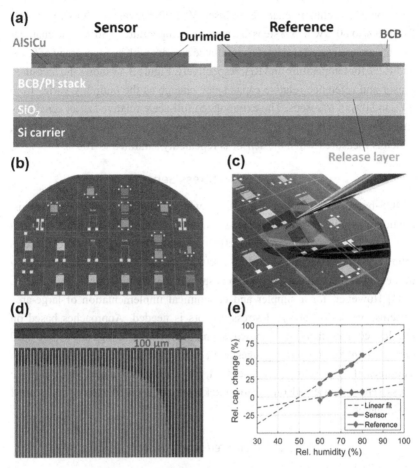

Figure 3.4 (a) Schematic of the foil substrate with the capacitive sensor and reference. The sensing material is Durimide (thickness about 2.7 μm) [99]. (b) and (c) Photographs illustrate the on-foil relative humidity sensors' samples before and during the mechanical release from the carrier wafer. [100] (d) A micrograph of the interdigitated structure of the capacitive humidity sensor. (e) Sensor and reference capacitors characterization results using LCR meter and climate chamber.

Figure 3.4c shows the simple mechanical release of the sensors from the carrier substrate. After the sensors are released from the carrier wafer, they are attached to printed circuit boards (both rigid and flexible), on which copper contact pads are predefined. The electrical connections from the sensor to the test PCB are made using silver glue. Sensors are then connected to a LCR meter for electrical impedance characterization.

A controlled climate chamber (Vötsch VCL0010) was used to vary the RH from 45% to 80% with 5% steps at a constant temperature of 30°C. The chamber sensors are calibrated and used as reference sensors with an inaccuracy of 0.6 K and 1.7% for temperature and RH, respectively. Figure 3.4e shows the measured sensor and reference relative capacitance change as the humidity varies from 60% to 80% in 5% steps. The extrapolated full-scale relative sensor sensitivity ($\Delta C/C_{RH=50\%}$) is about $\pm90\%$. The sensor response time is less than 6 seconds, which is measured when the sensor is repeatedly saturated to 100%.

3.2 Ultrathin Stress Sensors

CMOS-based stress sensors have been widely used in several applications, including robotic e-skin,[101] stress mapping for orthodontic brackets,[102] and compensation of piezo-Hall effect in Hall sensors.[103] The sensor's robustness coupled with the readout circuit precision enables accurate mapping of the object mechanical deformation especially when ultrathin chips are used. [104] However, for a simpler and economical implementation of large-area systems, an on-foil array of stress sensors is needed. Approaches based on interlocked nanofibers, Ag nanowires, and diffraction gratings are widely used to implement flexible strain gauges on PDMS substrates.[105], [106] Here, two examples of stress sensors are discussed, namely an integrated CMOS in-plane stress sensor fabricated on 20 µm-thick chips and on-foil silver-ink printed strain-gauge.

3.2.1 Integrated Stress Sensor

In [101], a HySiF demonstration is presented, which includes an in-plane stress sensor for monitoring the bending activity of a robotic gripper. PMOS and NMOS field-effect transistors are used to measure in-plane shear stress (σ_{12}) and normal stress difference ($\sigma_{11} - \sigma_{22}$), respectively. Figure 3.5a illustrates the sensor structure, where the difference between the current flowing in two orthogonal PMOS FETs is used to extract the in-plane shear stress (σ_{12}), assuming the piezoresistive coefficient of silicon π_{44} is given. Additionally, the current difference of two orthogonal NMOS FETs is used to extract normal stress difference ($\sigma_{11} - \sigma_{22}$), assuming the piezoresistive coefficients of silicon π_{11} and π_{12} are given. Designed for the stress range from -200 MPa (compressive) to $+400$ MPa (tensile) stress, the measured sensitivities are 17 nA/MPa and 12 nA/MPa for PMOS and NMOS arrangements, respectively. The sensors are fabricated using the 0.5 µm CMOS GFQ technology,[33] where the final chip thickness is about 20 µm.

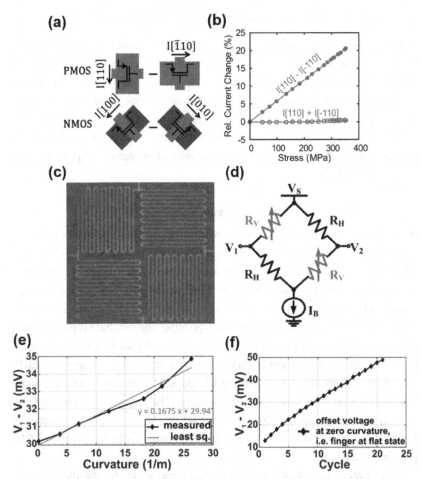

Figure 3.5 (a) In-plane CMOS integrated stress sensor. [24], [101] The device layout maximizes the sensor sensitivity for the in-plane stress components. (b) Measured relative current change of PMOS devices vs. the magnitude of the applied stress. (c) Photograph of four meanders printed using silver ink on BCB/PI substrate. [107], [25] The Wheatstone bridge circuit is utilized, in which two strain gauges are printed vertically (R_V), while the other two are printed horizontally (R_H). The line width is 20 μm and the total dimension of the bridge circuit is 5×5 mm^2. (d) A schematic of the Wheatstone bridge circuit, which has four resistors arranged in a diamond shape. The circuit has two current nodes, where the supply voltage V_S is connected to one node, while the biasing current I_B is connected to the other node. The difference between the two output nodes V_1 and V_2 is related to the change in the resistance values, as indicated using Eq. 3.1. (e) Strain gauge measurement results showing curvature of the gripper finger plotted against the output voltage of the strain gauge bridge circuit.[36], [37] Six measurements are taken during decreasing the bending radius of the gripper finger at bending radii of about 270, 142, 82, 55, 47, and 38 mm. (f) The offset voltage increases after each bending cycle.

3.2.2 On-Foil Bending Sensor

The principle of using the change in the electrical resistance to measure the strain distribution has been known for more than a century. This concept is widely used in different fields, such as biomedical applications, where an implantable strain gauge is used to monitor human bone strain or in aerospace applications,[108] where thin-film strain gauges are used to measure the thrust produced by a plasma thruster.[109] This principle is utilized here, where strain gauges serve as monitors of the bending activity of the robotic gripper adaptive fingers. [107], [22]

Aerosol Jet® Printing

Aerosol Jet® technique, also known as maskless mesoscale material deposition (M^3D), is utilized for this purpose.[110], [111] In this technology, the silver ink is atomized and transformed in a dense aerosol. The aerosol is transported to the printhead using an inert carrier gas. The printhead forms a high-velocity aerosol jet, which is capable of creating a minimum feature size of 10 µm.[110] During Aerosol Jet® printing, the substrate is placed on an XY stage and the printing velocities are set to 4–5 mm/s. A nozzle with an orifice diameter of 150 µm is used to print a line width of approximately 20 µm. A certain number of printing overpasses are performed on the same line in order to ensure reliable conductivity. The printing velocity and the number of overpasses determine the value of the strain gauge resistance. Figure 3.5c shows four strain gauges printed using this technique on a BCB/ polyimide substrate.

This printing technique has a main drawback, given by the difference in printing in horizontal and vertical directions. Silver droplets accumulate on the nozzle tip of the printhead during the printing, thus the aerosol jet shape becomes elliptical. This results in variations in the silver ink's density while printing in vertical and horizontal directions. Consequently, a mismatch takes place between the resistance values of the vertically and horizontally printed meanders.

Here, this drawback is minimized using a high number of overpasses, leading to a thicker metal layer. After each printing cycle, the resistance of each meander structure is measured using four-point measurements. Another layer of silver ink is deposited on the meanders with the higher resistance until the variations between the meanders are minimized. The minimum resistance variation reached is about ±5%. This mismatch between the strain gauge resistances results in an offset voltage between the output terminals of the strain gauge bridge circuit.

Strain Gauge Design

Figure 3.5c shows the meander or the zigzag design of the strain gauges, which takes advantage of the increased effective length of the silver ink. This design multiplies the effect of small strain variations, which occurs in the same direction as the printed silver-ink lines. [112] The strain gauge is more sensitive to the strain applied parallel to the direction of the printed lines and less sensitive (almost insensitive) to the strain applied in the direction perpendicular to the printed lines.

The Wheatstone bridge circuits are proven to be very effective implementations to accurately detect resistance variations while suppressing temperature effects. As shown in Figure 3.5d, the bridge circuit has two output terminals, V_1 and V_2, which are equal when the bridge is balanced. The bridge is balanced when the ratio of the two resistors in one branch is equal to the ratio of the two resistors in the other branch. One or more resistors can serve as a sensing active element depending on the type of strain and the required accuracy of the measurements. In order to achieve high bridge sensitivity and linearity, current-biased full-bridge topology is adapted in which two resistors (R_V in Figure 3.5d) are printed vertically to sense the bending strain variations, while the other two resistors R_H are printed horizontally in the direction perpendicular to the bending strain. Equation 3.1 relates R_V and R_H to the voltage difference between the output terminals V_{out}, given constant biasing current I_B:

$$V_{out} = V_1 - V_2 = I_B \cdot \frac{R_{V,1} \cdot R_{V,2} - R_{H,1} \cdot R_{H,2}}{R_{V,1} + R_{V,2} + R_{H,1} + R_{H,2}}. \tag{3.1}$$

The Wheatstone bridge is biased using a constant current source I_B as shown in Figure 3.5d. This biasing provides no measurement errors due to wiring resistance and a linear bridge response. It is important to note that the stability of the biasing current is crucial for the measurement accuracy, as any change in the biasing current is misinterpreted by the bridge circuit as a change in the resistance of the active elements. The linearity of the bridge can be illustrated, assuming $R_V = R_{V,1} = R_{V,2}$ and $R_H = R_{H,1} = R_{H,2}$ and using Eq. 3.1, which results in the following relation (Eq. 3.2):

$$V_{out} = \frac{I_B}{2} \cdot \Delta R_V + V_{offset}, \tag{3.2}$$

where ΔR is the change in the active resistors due to bending strain. The offset voltage V_{offset} corresponds to the mismatch between the two printing directions and is given:

$$V_{offset} = \frac{I_B}{2} \cdot (R_V - R_H).$$

Strain Gauge Characterization

An experimental study is performed on the printed strain gauge under tensile bending strain.[107] As the surface of the robotic finger experiences a nonuniform strain distribution during its bending activities, here the bending curvature is assumed to be the same within the relatively small area of each strain gauge (5×5 mm^2). The measurement setup consists of a column with a movable rod that is used to bend the strain gauge sample, which is mounted on the gripper finger material. Additionally, the side-view images are captured using a high-resolution digital camera. Consequently, the bending radius at the strain gauge position is extracted from the images by defining three points on the gripper finger curved surface and using a 5×5 mm^2 square grid as a reference. The output voltage of the strain gauge bridge circuit ideally varies linearly with the bending curvature given by Eq. 3.2.

Figure 3.5e shows the strain gauge output voltage plotted against the curvature of the gripper finger.[107] The full-scale output voltage change is about 5 mV during increasing the bending curvature at 1 mA biasing current. At the maximum bending curvature, the sensitivity of the strain gauge is evaluated by dividing the full-scale output voltage change by the total voltage drop on the strain gauge, which equals about 440 mV. Consequently, the sensitivity for the measured strain gauge is about 10 mV/V. At the flat state, the output voltage of the strain gauge is at the offset voltage, which traces back to the level variations of the individual bridge resistors.

The offset voltage V_{offset} is about 30 mV at a biasing current of 1 mA. The drift in the offset voltage after each bending cycle is shown in Figure 3.5f, which originates from the mismatch in the resistance between the vertical and horizontal meanders of the strain gauge. Here, the offset voltage increases with the increase in the resistance of the vertically printed meander, which is more sensitive to strain variations on the surface of the gripper finger. Since the gripper finger deforms after each bending cycle, the strain on the surface of the gripper finger does not equal zero at the flat position. Instead, it increases after each bending cycle due to the gripper material memory effect. Table 3.3 compares the Aerosol Jet®–printed strain gauge with other silver-ink printed gauges fabricated using screen and inkjet printing technologies.

3.3 On-Foil Antennas

Wireless data and power transmission features are in increasing demand in the era of connected things. Mechanically flexible sensor systems require antennas

Table 3.3 Comparison between printed stain gauges using silver ink.

Parameter	This work [107]	[169]	[170]
Printing Technology	Aerosol Jet®	Screen	Inkjet
Substrate	BCB/PI	Polyester resin	Kapton
Dimension (mm²)	5 × 5	9.3 × 4.6	10 × 15
Trace width (µm)	20	150	1000
Thickness (µm)	1–3	7	-
Sensitivity $\Delta R/R$ (%)	2	0.061	8
Gauge factor G	2	1.22	6.2
Sensitivity (mV/V)	10	100	37

that are mechanically and electrically tolerant to the levels of bending and stretching experienced in the targeted application. Therefore, simple lower-frequency antennas for RFID tags were the first to become commercially available. Driven by continuous improvements in the flexible substrate materials and metal printing technologies, the flexible antenna performance is steadily approaching that of those fabricated on conventional substrates.

An example is an inkjet-printed multiband antenna, which is fabricated on a Kapton substrate with a near-constant performance during bending.[39] Another example is a stretchable 915 MHz antenna that is fabricated by integrating conductive fibers in a PDMS substrate.[113] Note, the impact of the substrate surface topography, dielectric loss tangent on the conductive metal properties, and the overall antenna performance. In this section, 5.5 GHz and 13.56 MHz antennas fabricated on the flexible BCB/PI substrate are discussed. [114], [115]

3.3.1 Flexible 5.5 GHz Antenna

An ultrathin and conformal CFP-compatible antenna targeting the 5.5 GHz frequency has been fabricated on a flexible BCB/PI substrate.[45] The low loss tangent ($\tan\delta = 0.0008$ at 1 GHz) and low electrical permittivity ($\varepsilon r = 2.65$) of BCB, with its stable dielectric properties over a wide range of frequencies, is very attractive for many CMOS-compatible fabrication processes.[116] BCB is inherently available in the presented HySiF substrate, and its smooth surface allows for the fabrication of high-quality metal layers.

Figure 3.6a depicts the chip-embedding material stack in addition to the flexible antenna fabricated using AlSiCu metallization. During the fabrication process,[117] 3.5 µm and 10 µm thick BCB and PI are spin-coated and cured repeatedly. A thin layer of BCB (3.5 µm) is then spin-coated and cured, where

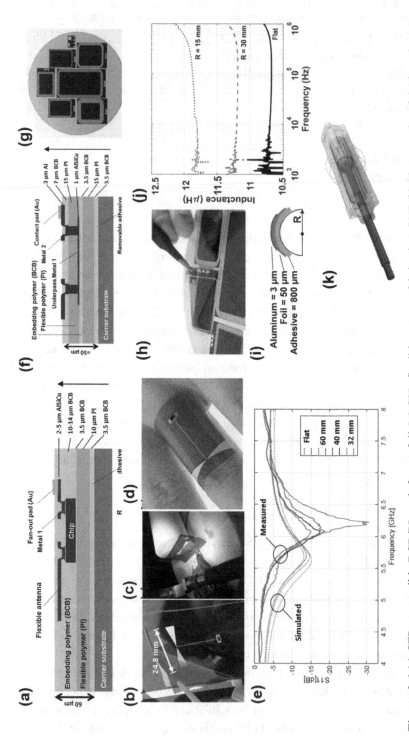

Figure 3.6 (a) CFP-compatible BCB/PI stack for on-foil 5.5 GHz flexible antenna fabrication. [55] (b) Photographs of the bowtie antenna while being released from the carrier wafer, (c) during S-parameters characterization using VNA, (d) demonstrating a wireless flexible

Caption for Figure 3.6 (cont.)

system. (e) Simulated and measured S11 of the flexible antenna at different bending radii of 60, 40, and 32 mm in addition to the flat position. [117], [118] (f) CFP-compatible BCB/PI stack for on-foil inductors fabrication. (g) Various loop designs, having different area, turns, and trace width and spacing, are integrated on a 150 mm carrier wafer. (h) Mechanical release of the inductors from the carrier wafer. (i) Illustration of the foil stack during bending, while highlighting each layer thickness. (j) Electrical characterization using LCR meter at different bending radii.[119], [120] (k) Robotic piston with integrated permeant magnet, courtesy of the company Festo AG.[120] This demonstrates another use case for the flexible spiral inductors in which an array of spiral inductors is placed on the piston and is used for energy harvesting. This design was done by Hahn Schickard and manufactured on Flex-PCBs by Würth GmbH & Co. KG within the ParsiFAL4.0 project.[120]

the cavity for the RF chip embedding is defined. A relatively thicker BCB layer is spin-coated that covers the chip surface. Vias for contacting the chip pads are opened by using BCB as a hard mask and plasma reactive ion etching. For realizing the chip fan-out and antenna metal layer, AlSiCu metal is sputtered and structured. Optional Au contact pads with an additional thin passivation layer of BCB could be added as the final fabrication step.

Figure 3.6b illustrates the mechanical release of the flexible antenna from the carrier wafer. The results of the S-parameters characterization are shown in Figure 3.6e for different bending radii. Here, a slight shift in the measured center frequency is observed, which can be attributed to the mismatch between the material properties used during FEM simulations and the actual BCB/AlSiCu properties.

3.3.2 On-Foil Inductor Loops

Passive inductors usually require large areas compared to other active or passive electronic components. Therefore, high-value inductors are avoided in VLSI filter and transceiver systems so that they can be integrated on the chip by economic means.[121] Small on-chip inductors are used in VLSI filter circuits, [122] while large off-chip inductors are inevitably used in wireless battery charging and NFC applications.[123], [124] Figure 3.6k illustrates another use case where an off-chip array of inductors is used to harvest energy from the movement of a permanent magnet core of a robotic piston.[120]

Figures 3.6f and 3.6g show the structure and an image of on-foil inductor loops fabricated on a BCB/PI flexible substrate primarily targeting NFC applications. A thin layer of AlSiCu metallization is used for the inductor loop underpass, while a thicker layer of aluminum (Al) metal is used to implement the main part of the loop for a designed trace resistance of less than 80 Ω. The loop design is compatible with the ISO/IEC 14443 standards' target smartphone applications. Table 3.4 lists the measured inductance for various loop sizes, turns, and trace width and spacing. Note that as the metal trace density increases, the stress induced on the flexible substrate increases, which leads to warpage and difficulties during the foil release from the carrier substrate. As a result, it is better to use fewer turns and a minimum trace width that satisfies the inductor quality factor requirements.

Figure 3.6h illustrates the simple mechanical release of the inductor loops from the carrier wafer. The loops are then mounted onto plastic cylinders, having different radii, using a foam adhesive as shown in Figure 3.6i. Figure 3.6j shows the frequency response of an on-foil loop (25×35 mm^2, 10 turns, trace width and spacing = 100 μm, trace thickness = 3 μm) at different

Table 3.4 Measured inductance of the fabricated on-foil inductor loops.[115]

Inductance (µH)	Resistance (Ω)	Size (mm²)	Turns	Trace width/ spacing (µm/µm)
3.6	9.2	41 × 55	5	600/300
5.2	16.6	41 × 71	5	400/400
3.3	10.5	35 × 40	5	400/200
4.4	41.3	25 × 35	5	100/100
13.8	81.3	25 × 35	10	100/100
7.7	20.1	35 × 40	10	400/200
11.6	27	35 × 40	15	400/200

bending radii.[115] The uniaxial tensile strain acting on the loop increases the inductor average diameter/area, hence the inductance value increases. For NFC communication, this inductance shift needs to be taken into consideration in circuit design.

3.4 Low-Voltage Organic Thin-Film Transistors

HySiF targets an economical implementation of ultrathin sensor systems by integrating a minimum number of embedded silicon chips and a maximum number of on-foil components. In large-area sensor systems, either many sensor readout channels or fewer time-multiplexed readout channels are implemented, requiring either larger area for the former case or higher number of input chip pads for the latter case. However, the sensor count could be extended while using the same silicon chip if an off-chip/on-foil multiplexer is implemented.

As an example, in [125], an array of piezoresistive strain sensors is multiplexed using p-n diodes fabricated on a plastic substrate. The sensors are arranged in the Wheatstone bridge configuration, where each array is addressed using ±3 V bias lines. In [6], the number of accessible large-area sensors is significantly improved compared to the normal binary address-ing by utilizing the frequency-hopping scheme. This prototype is imple-mented using an array of ZnO TFT-based oscillators fabricated on a glass substrate. Although this method increases the number of accessible sen-sors, the demonstration is fabricated on a glass substrate in addition to passive components and data reconstruction being integrated using con-ventional PCBs.

Figure 3.7a shows the structure of the OTFTs fabricated on a flexible BCB/PI substrate.[126] For extracting the transistor parameters, various devices with

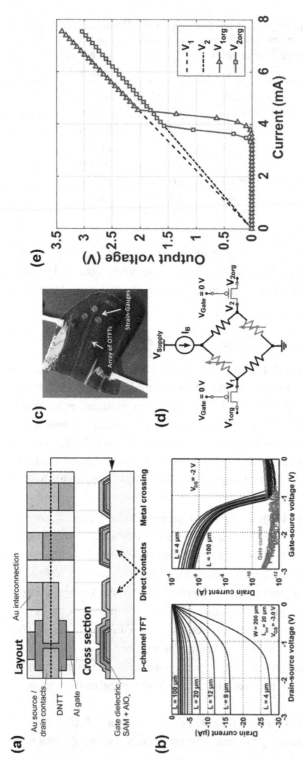

Figure 3.7 (a) Schematic showing the top and cross-sectional views of the organic TFTs and the circuit interconnects.[126] The gate electrode is aluminum (Al) and the source/drain contacts are gold (Au). The gate dielectric is a combination of AlO$_x$ and an alkylphosphonic acid self-assembled monolayer (SAM). The organic semiconductor is dinaphtho[2,3-b:20,30-f]thieno[3,2-b]thiophene (DNTT). (b) Measured static output and transfer characteristics of organic TFTs on BCB/PI substrate.[87] (c) Photograph of the OTFT array and three strain gauges fabricated on the flexible BCB/PI substrate.[87] (d) Schematic of a strain gauge connected to a pair of organic TFT switches.[86] (e) Static measurements showing that the organic TFT acts as a proper switch when VGS > |VTH|.

the same channel width (W = 200 µm) and different channel length (from L = 4 µm to L = 100 µm) are fabricated. Figure 3.7b shows the measured static output and transfer plots of the OTFTs. The threshold voltage, carrier mobility, and on/off current ratio are extracted, which equal –1 V, 1.3 cm^2/Vs, and 10^6, respectively. Here, the smooth surface of the BCB/PI substrate enables the OTFTs' performance to be comparable to that of those fabricated on glass.

For implementing an on-foil analog multiplexer, two arrays of OTFTs are fabricated on the BCB/PI substrate along with the on-foil bending strain gauge, previously discussed in Section 3.2.2. Figure 3.7c shows a photograph of the flexible plastic substrate with on-foil OTFTs and printed strain gauges. Figure 3.7e shows the static characteristics of the OTFT-based analog switch, which are evaluated by varying the bridge biasing current as illustrated in Figure 3.7d. As the biasing current increases, the bridge output voltages (V$_1$ and V$_2$ in Figure 3.7d) increase until the OTFT threshold voltage is surpassed, after which the OTFT turns on and a low-impedance path is created.

In order to properly design an OTFT-based integrated circuit and simulate its static and dynamic behaviors, the industry-standard BSIM3 model is modified and used to model the fabricated OTFTs.[126] This compact model is then used to create and simulate a low-voltage digital library, which includes inverters and NAND gates, and shift registers. Various logic families have been implemented, and for more details, refer to the corresponding implementation in [126]. Here, a shift register design is discussed, which is fabricated on the flexible substrate in order to automatically address the printed strain gauges. Figure 3.8 shows a one-stage shift register based on a dynamic positive edge-triggered master-slave flip-flop using the biased-load inverter design. The simulated and measured results of a three-stage shift register are shown in Figure 3.8c, which is performed using a supply voltage of 3 V, a bias voltage of –1 V, and a clock frequency of 100 Hz that can be extended to a record frequency of 3 kHz.

4 Ultrathin Chip Characterization

Semiconductor devices and integrated chips are always manufactured on rigid substrates for good reasons, as discussed in Sections 1 and 2. In this section, the impact of the thinning of the rigid substrate, with more than 90% material removal, on the integrated devices is addressed. This review includes electrical, thermal and mechanical characterization of devices, such as MOSFETs and BJTs, as well as functional integrated circuits, such as microcontrollers and A/D converters.

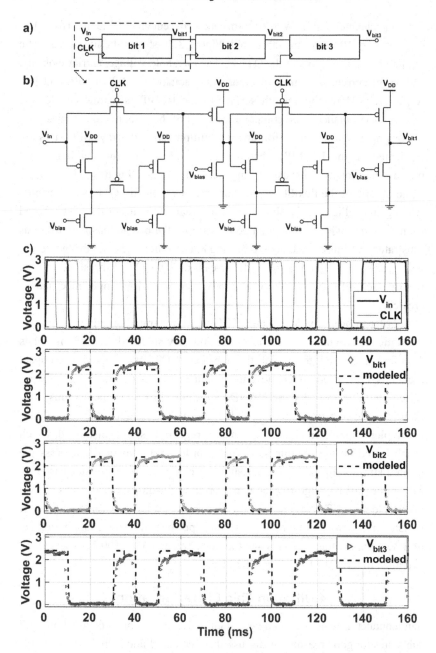

Figure 3.8 (a) Schematic of a three-stage shift register. (b) Schematic of a one-stage shift register using the biased-load inverter design. (c) Simulated and measured output voltage waveform of the three-stage shift register.[126]

4.1 Piezoresistive Effect in Silicon

Bending, stretching, compressing, and twisting are expected mechanical movements in a certain flexible electronic application. Fortunately, or unfortunately, depending on the targeted application, the natural electrical resistivity of silicon changes upon mechanical deformation of the semiconductor band structure and, thus, is directional. This electro-mechanical property is called the piezoresistive effect. One direct contributor to improving the overall speed of semiconductor devices is by engineering the crystal lattice strain strength and direction for maximum carrier mobilities.[127] Table 4.1 compares the effect of different stressors on n-channel and p-channel MOSFETs in the so-called strained-silicon technology.

By knowing the material resistivity ρ and the structure geometry, the resistance value R of a rectangular bar can be evaluated using Equation (4.1):

$$R = \rho \cdot l/(w \cdot h), \tag{4.1}$$

where the cross-sectional area of that bar is the width w multiplied by the thickness h. The current is assumed to flow along length l. Upon mechanical deformation of this bar, two types of effects take place, the macroscopic and the microscopic changes to the crystalline structure. The macroscopic change accounts for the geometric changes in length dl, width dw, and thickness dh of the bar. The microscopic change results in a variation of the material resistivity $d\rho$. Equation (4.2) can be used to express the combined effects:

$$dR/R = d\rho/\rho + dl/l - dw/w - dh/h \tag{4.2}$$

Table 4.1 Summary of the preferred strain direction for NMOS and PMOS transistors.

Strain Direction	NMOS	PMOS
Longitudinal	Tension (+++)	Compression (++++)
Transversal	Tension (++)	Tension (—)
Out-of-plane	Compression (++++)	Tension (+)
Preferred biaxial stressors	[110]	[110]
Maximum drain current enhancement exploiting all stressor options (including those not shown)	Up to 20%	Up to 70%

Having close values of Poisson's ratio, metal and semiconductor experience similar macroscopic geometrical changes upon mechanical deformation. However, the coefficient of piezoresistivity in metals is very small and can be neglected in most cases, leaving only the geometry variation factor to be evaluated. In contrast, for semiconductors, silicon in particular, the piezoresistive effect is far more pronounced than the geometrical change. The coefficient of piezoresistivity is a multidimensional tensor reflecting the change of resistivity with applied stress in different crystalline directions. It is important to consider the impact of the silicon substrate thinning, warpage, and packaging-induced offset strains on the piezoresistive properties of the ultrathin chip. [128] The thinner the substrate is, the more bending the chip can withstand, as indicated by Equation (4.3):

$$\sigma = \frac{E_{Si} \cdot h_{Si}}{2R}, \tag{4.3}$$

where σ is the resulting stress, E_{Si} is Young's modulus, h_{Si} is the substrate thickness, and R is the curvature radius. Even if uniaxial stress is applied to a flexible ultrathin chip package, the embedded chip experiences a highly nonlinear biaxial stress depending primarily on the mismatch between the chip and foil package Poisson ratios.[129] Stoney's formula is then used to calculate the stress resulting from that two-material system as shown in Equation (4.4):[130]

$$\sigma = \frac{E_{Si} \cdot h_{Si}^2}{6(1 - \upsilon) \cdot h_f \cdot R}, \tag{4.4}$$

where υ is Poisson's ratio, h_f is the foil thickness, and R is the curvature radius of the system. Additional process-induced stress results from the mismatch of the thermal expansion coefficients of the thin chip, package materials, and metal interconnects.

The piezoresistive effect in silicon enables the integration of multiple on-chip sensors with minimum changes to the IC fabrication process. As will be discussed in Section 5, MOSFET-based stress sensors can directly be realized by monitoring the current change in the transistor channel upon mechanical deformation.[101] Other monolithic integrated sensors, including pressure and flow sensors, utilize that principle.[131], [132] However, for these compact sensor SoCs to be mechanically flexible, the silicon chip thickness should be less than 50 μm.[133] When the effect of mechanical deformation on different blocks of such ultrathin SoCs is carefully considered, a conflict of interest arises between the transducer and the sensor readout electronics. On the one side, the sensor's sensitivity almost always benefits from larger mechanical deformation. However, the precision of the readout electronics, the analog part in particular, could be negatively impacted.

4.2 Integrated Devices

Transistor scaling, being the key concept for improving integration density and transistor performance, might soon be approaching its limit. Strained-silicon technology has steadily allowed for further improving device density and speed.[127] Therefore, extensive work has already been done in characterizing the stress/strain effects in silicon on monolithic active and passive devices. Thinning silicon substrates down to the microscale range of 5–50 μm will likely not impact the silicon electrical properties compared to bulk silicon substrates.[134] Only for nanoscale substrate thickness (< 100 nm) do silicon thin-film properties take over, as fundamental material constants like the Poisson ratio, Young's modulus, and electrical permittivity become thickness-dependent variables. Therefore, bulklike device and circuit design can be pursued in spite of thinning chips to the 5–50 μm range.

In the following section, a review of stress effects on monolithic resistors, capacitors, and transistors is presented. The impact of the nonideal chip grinding techniques on the electrical performance of electronic components, when integrated into microscale silicon substrates, also is discussed.

4.2.1 Resistors

Resistors are essential components for any integrated circuit. In modern CMOS IC technologies, different resistor types are available. Table 4.2 lists such integrated resistors and compares their stress dependencies.[135] Metal resistors typically have low sheet resistance and, thus, are less used due to the consequently large chip area and power consumption. In contrast, implanted and diffused silicon resistors are economical since they are readily available in the standard CMOS fabrication process and can be made small and compact, since they can be adjusted to sufficiently high sheet resistivities. However, their stress sensitivity is large, since their sheet resistance is determined based on the dopant diffusion profile in silicon. Figure 4.1 shows the measured resistivity plot at different bending radii for n- and p-type highly doped silicon diffusion resistors fabricated on bulk silicon and 20 μm chips attached to Kapton foils.

Table 4.2 Comparison of resistor types available in modern IC technologies. [135]

Resistor type	Metal	Diffusion	Poly	Silicided diffusion	Silicided poly
Sheet resistance	Very small	Large	Large	Medium	Medium
Stress sensitivity	Small	Large	Medium	Small	Small

Polysilicon resistors are more commonly used, since they provide an optimized solution for area and stress sensitivity. However, when targeting mechanically flexible systems, silicided diffused single-crystalline silicon and silicided polysilicon resistors should also be considered, as they provide the least stress sensitivity and a clear hybrid combination of metal and polysilicon or diffusion resistors. Consequently, the chip area will be traded against stress sensitivity in this case. Different layout techniques can also be used to minimize stress sensitivity, such as orthogonal resistors in a Wheatstone bridge or horseshoe layout.

4.2.2 Capacitors

Different types of capacitors are available in modern semiconductor technologies. Metal-oxide-semiconductor (MOS) parallel-plate capacitors are readily available in CMOS technologies. They are used when high-density capacitance per area is needed as in IC power supply filtering or as a voltage-controlled capacitor (varactor). Obviously, MOS capacitors have the same stress dependency as MOSFETs. MOS capacitors are replaced by metal-insulator-metal (MIM) capacitors in precision circuits due to their superior linearity. Coupled with FEM modeling, planar fringe capacitors are increasingly used in DSM technologies, as they provide low capacitance values suitable for ultralow-power applications.[136]

Figure 4.2a illustrates the effect of stress on the parallel plate structure. The parallel plates of the MOS and MIM capacitors are commonly placed in at least two different metallization layers (more layers are needed in the case of triple MIM capacitors) or one metal layer and the silicon substrate, respectively. Both plates exhibit geometric changes upon stress variations, which lead to a slight increase in the capacitor surface area and capacitance

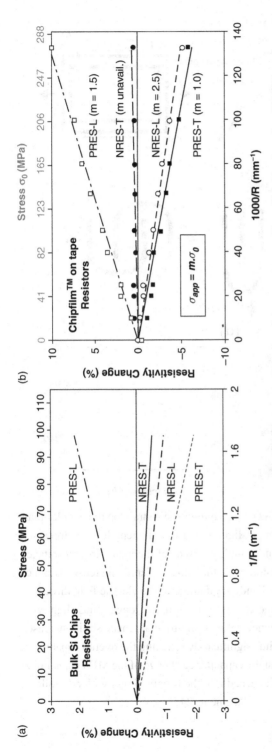

Figure 4.1 Measured normalized resistivity change plotted against the inverse of the curvature radius of highly doped n-(NRES) and p-type (PRES) silicon resistors.[129] (a) Thick bulk silicon and (b) 20 μm chips glued and wire-bonded to Kapton foil.

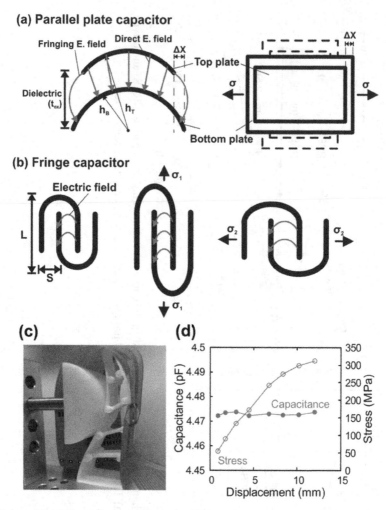

Figure 4.2 (a) Exaggerated effect of applying tensile strain on the parallel-plate capacitor. Upon tensile strain, a slight increase in the length of the top and bottom plates. Due to the high Young's modulus of the metal, the capacitance variation is negligible. (b) Illustration of the impact of applying tensile strain on the fringe capacitor transversal and longitudinal to the electric field direction.

Transversal tensile strain increases the metal finger length, thus slightly increasing the overall capacitance value. Longitudinal tensile strain increases the finger dielectric spacing, thus significantly reducing the overall capacitance value. (d) Stress plotted against the capacitance change of the MIM capacitor as an object, as shown in (c), is pushed into the robotic finger with controlled displacement.[24]

value upon tensile strain. The parallel-plate capacitance change is then given by:

$$\Delta C = \frac{\epsilon\left(L_0 + dL\right) \cdot \left(W_0 - dW\right)}{t_{ox}} - \frac{\epsilon\left(L_0 \cdot W_0\right)}{t_{ox}},$$

where L_0 and W_0 are the initial capacitor length and width, and dL and dW are the corresponding incremental geometric changes and are related using Poisson's ratio. t_{ox} is the dielectric thickness, which is assumed to be constant. Following Eq. (4.2) and (4.3), the resulting stress is directly proportional to the distance from the neutral plane of stress, which resides near the center of the silicon chip body. Therefore, the top plate experiences slightly higher stress than does the bottom plate. Note that this effect is negligible for a chip having conventional silicon thickness but may be noticeable in UTCs. Due to the high Young's modulus of the metal plates, the overall capacitance variation due to stress is minimal and negligible. Note that other effects could take place when different shear stresses are applied to the top and bottom plates or when the dielectric material properties are very different from those of silicon. It is worth mentioning that few dielectrics have shown stress-dependent electrical permittivity.[136], [137]

Unlike the parallel-plate capacitor, the fringe capacitor is strongly sensitive to stress variations. Figure 4.2b shows the effect of tensile strain when the stress is applied transversal or longitudinal to the electric field. For the transversal stress, the finger length is varied slightly depending on the metal's elasticity, while the finger space is almost unchanged. However, for the longitudinal stress, the fingers' spacing is altered depending on the dielectric material's elasticity. Therefore, the fringe capacitors are more sensitive to longitudinal stress, since dielectric materials have in general lower Young's modulus than the metal electrode.

Figure 4.2c shows a Kapton foil attached to a robotic finger, which leads to bending tensile strain on the outer surface of the finger. A 20 µm chip is glued to the Kapton foil, and the capacitance variation of an on-chip MIM parallel-plate capacitor is recorded. No stress sensitivity is observed in the measured capacitance level, which could be limited by the accuracy limit of the LCR meter.

4.2.3 Diodes and Transistors

MOSFETs connected in diode configuration are used in electrostatic discharge (ESD) protection circuits, which are placed near the chip pads for I/O, supply, and ground connections. Here, such ESD devices are used to perform continuity tests in order to check the functionality of the ultrathin chips at different processing stages. The continuity test is important in the case of ultrathin chip embedding using CFP technology, where fine-pitch interconnects are used to connect the embedded thin chip to the on-foil fan-out pads.

Figure 4.3 shows the measurement results of the ESD structures integrated into the Apollo microcontroller.[138] For the first test, the digital ground voltage V_{SS} is connected to ground and the digital supply voltage V_{DD} is swept from 0 to –0.5 V using an SMU. Since multiple diode-connected p- and/or n-channel MOSFETs are available at ground, supply, and I/O pads and are connected in parallel between V_{DD} and V_{SS}, the measured current for this first test is the total current of all connected ESD devices. Note that for each ESD circuit, p- and n-channel MOSFETs are connected in series. This test checks the overall state of the simple integrated ESD devices but does not say much about the microcontroller performance or the individual I/O pads.

Figure 4.3b shows four sets of measured diode characteristics for 400 μm and 30 μm chips glue-attached to an LCC ceramic package, 30 μm chips during CFP processing after metal 1 structuring, and the same 30 μm chips after finalizing the CFP processing. From Figures 4.3c and 4.3d, the chips from the CFP show higher leakage currents and lower on-currents that can trace back to higher mechanical stress levels and temperature present during the CFP fabrication steps. Note that there are slightly lower on-currents for the 30 μm chip in a ceramic package when compared to the 400 μm chip in the same package, which results from the defects introduced by the back-thinning process.

For the second test, V_{SS} and V_{DD} are connected to 0 V and 2.5 V, respectively. Additionally, the voltage of one I/O pad (PAD7) is swept from –0.8 V to 3.3 V. In this way, both diode-connected devices on this I/O pad are separately characterized. Furthermore, the ultrathin chip embedding using the CFP fabrication process is electrically validated for each individual I/O pad.

Figures 4.3e and 4.3f show the same four sets of diode characteristics plot for the ESD n-channel and p-channel MOSFETs but for one I/O pad now. For the n-channel devices, the current curves are the same after deposition of Metal 1 and after the completed CFP processing. However, for the p-channel devices, the current increases after the CFP processing, which could indicate a slight stress relief on the embedded chips in the preferred direction of the p-channel devices.

4.3 Ultrathin Microcontrollers

Low-power microcontroller ICs (μCs) are at the heart of all high-performance flexible electronic applications. The Apollo low-power microcontroller,

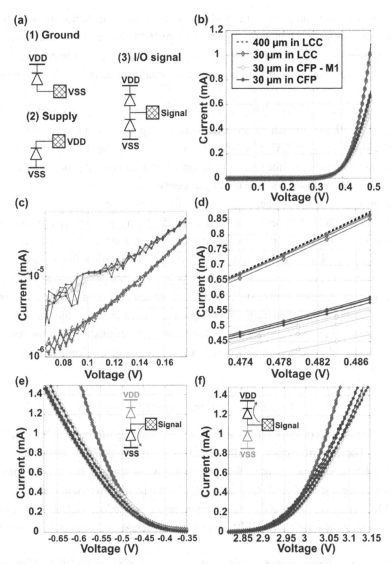

Figure 4.3 (a) Basic ESD configurations for ground, supply, and I/O chip pads. (b) Diode characteristics divided into four sets, each containing 4x chip samples. The first set corresponds to 400 μm chips glued to standard ceramic package. The second set contains 30 μm chips in the same package. The third set is for the chip characterization during the CFP processing. The fourth is the characterization results after finalizing the CFP fabrication. (c) and (d) are zoomed curves for the off- and on-diode currents. (e) and (f) are the diode characteristics of ESD n-channel and p-channel MOSFETs at one I/O pad of the Apollo microcontroller.

designed by the company Ambiq Micro,[138] is used for demonstration and evaluation purposes. It includes a powerful main CPU and multi-standard wired communication modules. However, this μC is not designed while considering external mechanical deformation, which is usually experienced by ultrathin chips.

Back-thinning of bare dies is used here to thin the μC ICs to a thickness of less than 30 μm, which is an intermediate step toward the ultrathin chip embedding using the CFP fabrication process. For the CFP verification, a method is developed for testing ultrathin chips after critical processing steps such as back-thinning and fine metal pad connections.[139]

Figure 4.4a shows a schematic diagram for the developed method, in which the ultrathin μC is the device under test (DUT). Samples of the thin chips are glued into standard IC packages, and the chip pads are wire-bonded to the package pins. Note that dealing with bare-die μC ICs requires the manufacturer's evaluation board (EVB) to be able to debug and flash the external μC ICs. Using an independent power supply is recommended to decouple the DUT from the test setup. Simple programs are designed to test different μC blocks such as the integrated temperature sensor. In the following sections, various blocks of the μC are electrically and thermally characterized before and after back-thinning.

4.3.1 Clock Generation and Timing

The Apollo microcontroller IC includes low- and high-frequency RC oscillators. The low-frequency oscillator (LFRC), with a nominal frequency of 1,024 Hz, is commonly used for low timing accuracy and basic finite state machine operations.[138] However, the 24 MHz high-frequency oscillator (HFRC) is used when stable and accurate timing is needed, such as in SPI or I2C communication. The HFRC could be switched off during low-power operation.

In order to characterize the impact of the chip thinning on the RC oscillators, the period jitter of the LFRC and HFRC oscillators is measured. The period jitter is defined as the deviation in cycle time or period of a clock signal with respect to its ideal value over a number of randomly selected cycles.

The period jitter measurement results are performed by using a simple program, which enables the corresponding oscillator and connects its output clock signal to a specific general-purpose I/O (GPIO) pin. A high-speed oscilloscope is connected to the clock I/O pin, and then 2,000 successive clock cycles are recorded. The same experiment is repeated for three different chips, namely the μC EVB, 400 μm-thick, and 30 μm-thick bare die chips. The EVB

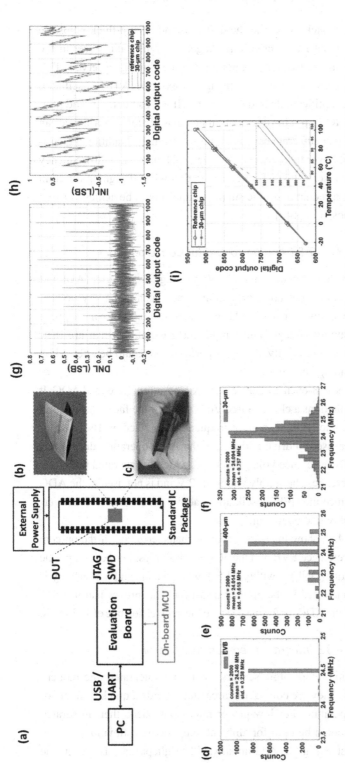

Figure 4.4 (a) Schematic showing the methodology developed for testing (b) ultrathin and (c) embedded μC ICs.[139] The test uses an evaluation board with on-board reference μC for test verification. The device under test (DUT) is connected to the evaluation board using a JTAG and/or SWD debugging interface. Period jitter of the on-chip RC oscillators. (d)–(f) correspond to the HFRC oscillators period jitter. (d) This is measured using the evaluation board (EVB) of the Apollo microcontroller, which uses a standard BGA package (e) for the 400 μm bare die mounted in LCC ceramic package, while (f) is of the 30 μm-thick dies. Static histogram-based (g) DNL and (h) INL of the on-chip 10-bit SAR ADC for the reference and 30-μm-thick chips. (i) 10-bit digital code of the on-chip temperature sensor plotted with the temperature of the chamber.[99]

uses a standard BGA package, while the 400 μm- and 30 μm-thick chips are placed in a conventional LCC ceramic package. A slight static offset in the 1,024 Hz LFRC oscillator frequency among the three chips is observed. This frequency mismatch can be trimmed using an external XTAL oscillator. Figure 4.4f shows a significant degradation in the HFRC oscillator period jitter for the 30 μm-thick μC chip. This degradation in the oscillator performance can be attributed to a higher substrate capacitive noise coupling. Traditional guard-rings are effective at lower frequencies; however, such unusual high-frequency noise effects should also be taken into consideration when designing ultrathin chips. For example, a reduction in the supply bounce could be achieved using the techniques proposed in [140].

4.3.2 10-Bit Analog-to-Digital Converter

Most modern μCs include a multichannel analog-to-digital converter (ADC). It is usually used for coarse detection of analog events such as battery voltage levels or setting temperature thresholds. Usually, analog circuits are more prone to PVT (process, supply voltage, and temperature) variations than are their digital counterparts. Consequently, the ADC performance could be affected by chip thinning or bending activities.

The Apollo μC has an 8-channel, 800 kSa/s SAR-based single-ended ADC. It uses the HFRC oscillator as a clock source. As a result of the facts discussed in the previous section, the HFRC oscillator frequency is divided by 16 for better accuracy and is then used to drive and trigger the ADC operation during this experiment. The ADC reference voltage is 1.5 V, which is generated internally using a bandgap reference. The supply voltage is 2 V and is not used as the ADC reference voltage, since the HFRC oscillator is active during the ADC conversion and causes supply line bouncing.

Figures 4.4g and 4.4h show the static linearity characterization of the on-chip 10-bit ADC for the reference EVB and the 30 μm-thick chips. For both chips, the differential nonlinearity (DNL) is within 1 LSB, while the integral nonlinearity is within ±1 LSB, as indicated in the product datasheet. This proves that the chip thinning did not alter the ADC performance at the low conversion rate of 125 kSa/s.

4.3.3 Integrated Temperature Sensor

On-chip low-precision temperature sensors are always integrated into modern microcontroller ICs. They are compact in area and provide a quick estimate of the silicon die temperature. For low-power microcontrollers, their integrated temperature sensor could be used for ambient temperature estimation, as chip self-heating is almost negligible. Additionally, if the right package is chosen, the condition of low thermal resistance between the silicon body and ambient air also is fulfilled.

The Apollo μC integrates a bandgap circuit temperature sensor. A conventional bandgap circuit has two voltage levels; one is the PTAT (proportional to absolute temperature) voltage level, and the other is the CTAT (complementary to absolute temperature) voltage level. The difference between these two voltage levels is ideally a temperature-independent voltage level.

A diode-connected MOSFET or BJT device is the main component of a bandgap reference circuit. It is worthwhile to mention the collector current I_C of a diode-connected ($V_{BC} = 0$) BJT device as a function of its base-emitter voltage V_{BE}, emitter area A_E, and reverse saturation current I'_S:

$$ I_C \mid_{V_{BC}=0} \approx I_S \cdot e^{\frac{V_{BE}}{kT/q}} . $$

The base-emitter voltage can thus be expressed as follows:

$$ V_{BE} = \frac{kT}{q} \cdot \ln\left[\frac{I_C}{I_S}\right], $$

where the difference between two V_{BE} voltages from two different devices provides a linear PTAT voltage. However, by examining the absolute value of the V_{BE} voltage and considering the process dependency of I_c and I_s, the following relation could be derived:

$$ V_{BE} = V_G - \frac{kT}{q} \cdot \ln\left[\frac{A_E \cdot K_1 \cdot T^\gamma}{I_C}\right], $$

$$ V_G \approx V_{G0} - a \cdot T, $$

where V_{G0} is the silicon bandgap energy extrapolated at 0 K, and a, K_1, and γ are process-dependent constants. The term V_G provides the CTAT voltage; however, V_{BE} includes an additional nonlinearly temperature-dependent term. The 30 μm-thick μC chip is assembled in a ceramic package. It is placed in a climate-controlled chamber, in which the temperature is varied from $-20°C$ to $100°C$ in $20°C$ steps and each step lasts for 2 hours. Figure 4.4i shows the 10-bit digital output of the integrated temperature sensor for the 30 μm-thick die and another reference chip. The data is captured by using a single-wire output (SWO) interface to a debugger chip on the EVB, which passes the sensor data to a PC using a USB interface.

By investigating the temperature sweep results, an offset is observed between the sensor readings of the reference and thinned chips. This offset can be trimmed. However, the sensor sensitivity, calculated as the slope of the straight lines in Figure 4.4i, slightly decreased from the nominal value of 3.6 mV/°C to 3.5 mV/°C. This can be regarded as the increase in the CTAT temperature-dependent leakage current of the bandgap reference circuit as defects in the silicon substrate approach the active area after chip thinning.

4.4 Ultrathin Capacitive Sensing Circuit

Different approaches for implementing capacitive sensing interface circuits are available in the literature. However, the choice boils down to whether the targeted capacitive sensor is integrated into the readout chip or the sensor is placed off-chip.[141] In the previous section, a HySiF-compatible on-foil humidity sensor was discussed, which will be tested in this section with a capacitive readout IC. The readout IC is back-thinned to a thickness of 30 µm and placed in a standard ceramic package. The sensor is then placed on an interposer foil and the sensor electrodes are connected to the interposer traces using silver glue. The sensor system is then placed in a climate-controlled chamber (Vötsch VCL 0010), in which the relative humidity (RH) is varied from 45% to 80% with 5% steps at a constant temperature of 30°C. The chamber sensors are used as reference sensors with an inaccuracy of 0.58 K and 1.7% for temperature and RH, respectively.

Here, the readout circuit comprises a programmable charge amplifier and an 8-bit ADC. The 5×5 mm^2 on-foil sensor and reference capacitors (nominal value about 30 pF) are arranged in a half-bridge configuration that is excited with two out-of-phase square wave signals.[99] During the CMOS chip design, expected stress levels inside the CFP package are considered in simulation as statistical variations such as for other process parameters.

Figure 4.5a presents the measurement results for the on-foil capacitive relative humidity sensor when connected to the 400 µm- and 30 µm-thick readout chips assembled in a ceramic package.[26] They show a similar response in terms of linearity, where the calculated coefficient of determination (R2) equals 0.9868 and 0.9867 for the 400 µm and 30 µm chips, respectively. Figure 4.5b shows the identical behavior of the sensor system during adsorption and desorption, which proves the reliability of the capacitive sensor and that it shows minimum hysteresis. For the dynamic characterization, pulses of human-exhaled breath were used to saturate the sensor with water vapor. Figure 4.5c shows the sensor dynamic response, in which the ADC digital code is plotted and the response times during adsorption and desorption (defined as the rise/fall

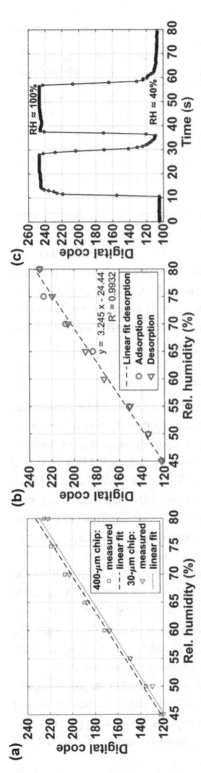

Figure 4.5 (a) and (b) Measured digital code of the 8-bit ADC with varying the RH in a controlled climate chamber.[26] (c) Real-time plot of the sensor system digital output upon contact with the exhaled human breath as it saturates the sensor with water vapor.[99]

Table 4.3 Comparison of different TFT technologies used to fabricate flexible ADCs

Technology	Si-UTC	Poly-Si	a-Si	OTFT	Metal-Oxide
Resolution	++	+	-	-	-
Speed	++	+	+	-	-
CMOS availability	++	++	-	+	-
Room-temperature processing	-	++	++	++	++
Large-area coverage roll-to-roll	-	+	++	++	++

time and calculated from 10% (90%) to 90% (10%) of the signal value) are about 3 and 6 seconds, respectively. Note that the sensor response time is limited here by the added circuit parasitics and readout ASIC. However, the intrinsic sensor response time can be much faster, which again favors the HySiF approach that brings the readout electronics as close as possible to the sensing element.

4.5 Ultrathin Flexible ADCs

The next generation of mechanically flexible RFID devices calls for miniaturization (i.e. more thickness reduction), higher computational power, and more integrated sensors and sensor readout channels while keeping power consumption minimal. ADCs are at the heart of every smart sensor. Although ADCs integrated into ultrathin chips represent the natural extension for the silicon-based semiconductor technology to the flexible electronics world, ADCs fabricated using other technologies, notably organic and inorganic metal-oxide semiconductors, are on the rise. These technologies allow for large-area coverage and are inherently flexible and simply processed in a room-temperature environment.

To this end, the Silicon-on-Polymer package has been developed by the company American Semiconductor in which the industry's first bendable ADC is embedded.[142] The 8-channel 8-bit 100 kS/s SAR ADC is fabricated using a 180-nm CMOS technology, and its die size is about 2.5×2.5 mm^2. The so-called FleX-ICs are essentially UTCs with thickness of 25 μm and are embedded in 50 μm-thick polymeric packages using spin-on polymers.

Alternatively, a 6-bit SAR ADC is fabricated on PI substrate using unipolar dual-gate metal-oxide (InGaZnO) TFTs.[90] The ADC occupies an area of 27.5 mm^2 and consumes about 73 μW when operated using a supply voltage of 15 V at a conversion rate of 26.67 S/s. Other examples include $\Delta\Sigma$-based ADCs fabricated on polymeric foils utilizing organic TFTs or InGaZnO TFTs,[143], [144] which achieve Walden FoM of 3.45 and 0.039 μJ/conv.-step, respectively. Table 4.3 compares different TFT technologies that are used to fabricate flexible ADCs, namely single-crystalline UTCs, polycrystalline silicon (poly-Si), amorphous silicon (a-Si), organic TFTs, and metal-oxide TFTs (e.g. InGaZnO).

5 Hybrid Systems-in-Foil Demonstrators

Although the Hybrid Systems-in-Foil (HySiF) concept is generic, ultrathin chip (UTC) manufacturing and packaging technologies, such as the ChipFilm Patch (CFP), are at the core of the presented HySiF realization. For the purpose of demonstrating the practical feasibility of the CFP embedding technology, various sensor systems have been demonstrated. In this section, two HySiF demonstrators are discussed; namely, an electronic skin (e-skin) for a bionic handling assistant, [145], [146], [147], [101], [25], [89] and a flexible smart label for activity tracking.[120] These demonstrators highlight the key merits of HySiF, in which the high-performance ultrathin ICs and the large-area passive and active components are complementary, integrated into and on flexible polymeric substrates.

5.1 E-Skin for Bionic Handling Assistant

Driven by the potential of decreasing the gap in the human–machine interactions, the company Festo AG has invented and pioneered the bionic handling assistant. [145] This achievement was awarded the German future prize in 2010. Figure 5.1a shows the 3D-printed resilient gripper arm of the system, the structure of which imitates an elephant's trunk. The arms, as well as the gripper fingers, are inherently flexible and able to hold and transport sensitive as well as irregular objects. In order to extend the bionic handling assistant functionality and awareness, the integration of electronic sensing components with the flexible body of the robot is required. However, most of the commercially available electronic products nowadays are mechanically inflexible and thus not suitable for seamless integration with the robotic body. This demonstrator has been developed under the framework of KoSiF project, which stands for "Komplexe Systeme in Folie."[146]

5.1.1 Integration Technology: ChipFilm Patch as Sensor System

Several challenges are tackled for the successful realization of the high-performance HySiF when different technologies are integrated on the same substrate. As has been mentioned in the previous sections, thin chips with a thickness of less than 30 μm have been successfully manufactured and packaged using the CFP embedding technology.[89] Furthermore, an array of strain gauges is printed using aerosol jet with silver inks on a BCB/PI substrate.

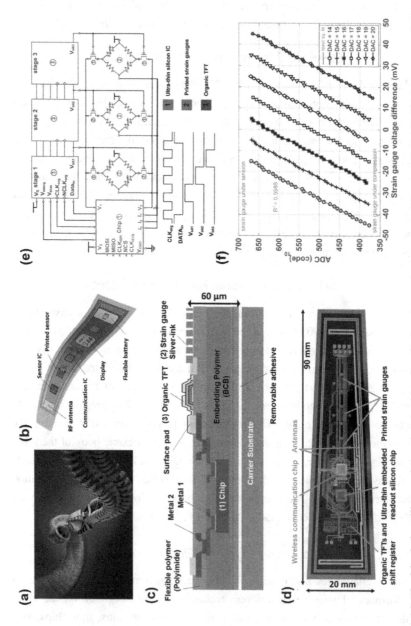

Figure 5.1 (a) The bionic handling assistant inspired by the elephant's trunk, courtesy of Festo AG.[145] (b) A conceptual illustration for HySiF comprising different flexible electronic components.[3] (c) Cross section of HySiF fabrication technology. Fabrication starts with CFP

Caption for Figure 5.1 (cont.)

ultrathin chip embedding, followed by strain gauge printing, and finally organic TFT fabrication.[5] (d) A photo illustrating the e-skin HySiF after its release from the carrier wafer.[5] The foil is then attached to the bionic handling assistant using a simple foam adhesive. (e) Block diagram of the e-skin HySiF and its timing diagram.[22] It includes printed strain gauges, organic shift register, OTFT analog switches, and an ultrathin silicon readout chip. A single 5 V supply is used to operate this system. (f) Static measurement results of the sensor readout circuit implemented in a 20 μm-thick silicon chip at different offset cancellation voltages.

[86] Additionally, a complete digital library based on low-voltage organic thin-film transistors (OTFT) is achieved.[126] However, what is lacking at this point is to implement the different components within or onto the single foil carrier, while maintaining their electrical performance and mechanical flexibility. By including all boundary conditions with respect to materials, surface properties, and temperature budget, successful HySiF integration is achieved.

Figure 5.1c shows the cross section of the e-skin foils. The HySiF fabrication starts by following the CFP embedding technology, in which the 20 μm-thick sensor readout IC is embedded in BCB/PI substrate in a face-up process. The CFP sandwich structure ensures the placement of the active layer of the chip near the neutral plane of stress of the foil carrier in order to minimize process-induced stresses on the UTC. Conventional lithography is used to structure two layers of AlSiCu interconnects, and the adaptive layout technique is used to compensate for the errors during the chip placement. The minimum thickness and homogenous density of the metallization safeguard against the warpage after the foil release from the carrier. A stress-relief layer is deposited on the carrier silicon wafer before the CFP, which compensates for the inherent bow of the silicon wafers.

After the final BCB/PI spin-coating, the wafer undergoes a plasma cleaning step in preparation for the strain gauge printing. Aerosol jet printing with silver ink is then used to manufacture the strain gauges on the smooth surface of the polymeric substrate. Finally, the OTFT-based circuit is fabricated in quasi-room-temperature conditions and by silicon stencil masks for achieving high-quality and high-resolution structures.

5.1.2 System Design

The demonstrator included three HySiF-based e-skins, each for one of the gripper's three fingers. The first foil is called the Smart Display, which includes a flexible display, an ultrathin display driver chip, and a stress sensor UTC. The second foil is the Smart Sense foil, which has four distributed stress sensor chips,[24] a wireless communication chip in addition to 27 MHz and 868 MHz antennas. The third foil and the main focus of this section is the Smart Ink Sense foil, which comprises the following components: (a) an array of printed strain gauges; (b) an organic switching circuit; (c) an ultrathin readout and wireless communication chips; (d) printed antennas and e) a thin-film battery.

Figure 5.1e shows the block diagram of the e-skin HySiF and its timing diagram. Here, the focus is on the strain gauges, readout chip, and organic circuitry. Note that the wireless communication chip, antennas, and thin-film battery will not be addressed. During the bending activity of the bionic handling assistant, the gripper fingers' outer surface experiences a stress distribution corresponding to the shape of the object caught between its fingers. This stress distribution is sensed by the three strain gauges, which develop a voltage difference between their output terminals.

The readout chip generates biasing voltages and currents for the organic addressing circuit and the strain gauges, respectively. The readout chip also sends a pulse signal with a duration of 1 ms to the input of the organic addressing circuit. The pulse propagates through the OTFT-based shift register and successively turns on a pair of p-type OTFT analog switches. The readout circuit amplifies and quantizes the stress-dependent voltage signals of each strain gauge. In order to minimize the circuit design footprint and its power consumption, the readout chip, as well as the strain gauges, operate using duty cycling, and they are switched off when the HySiF is sleep mode.

5.1.3 Operation

Figure 5.1d shows a photograph of the e-skin HySiF, highlighting the flexible electronic components. After the HySiF is released from the carrier wafer, the foil is attached to the bionic handling assistant using a simple double-sided foam adhesive. Basic configuration commands are sent from the communication chip to the readout chip. An example is a command required by the readout circuit to compensate for the mismatch in the sensor output voltage resulting from the nonuniform printing process. A 5-bit digital-to-analog converter (DAC) and a differential difference amplifier (DDA) are integrated into the readout chip and used for the sensor signal offset compensation and amplification.

Figure 5.1f shows the measured 10-bit ADC output of the readout circuit at different offset cancellation modes. Implemented in 0.5 μm CMOS, the readout chip consumes less than 2 mA from a single 5 V supply voltage. Good linearity is achieved, proving the effectiveness of the readout approach even though the chip is only 20 μm thick.

5.2 Flexible Smart Label for Activity Tracking

The first generation of the smart labels was enabled by the RFID technology, in which a low-cost RFID tag is implemented using a compact silicon chip with two pins to a printed antenna. The reader device takes over the interrogation task and provides an electromagnetic field strong enough to power the nearby tags. Driven by Moore's law and the rise of artificial intelligence, more and more computation power is implemented locally in the sensor nodes. Nowadays, the second generation of RFID tags integrates not only communication and memory blocks but also includes signal-processing cores, multichannel ADCs, support for various communication protocols, integrated sensors, and sensory peripherals for off-chip sensors.[148], [149], [150], [151]

Figure 5.2a shows a conceptual schematic of a high-performance mechanically flexible smart label for the purpose of logistics tracking. Here, several ultrathin sensors and microprocessor chips are integrated in order to provide a close-to-real-time monitoring of the product status in the journey from the supplier to the end-user. Accelerometers are used to detect sudden and unplanned collisions. Temperature, humidity, and pressure sensors are used to continuously monitor the environmental conditions outside and inside the package. Near-field communication (NFC) allows reliable label configuration at the manufacturer facility and secure payments and financial transactions at the end-user. A longer-range communication technology, Bluetooth or WLPAN as an example, connects multiple labels to a base station within the framework of the Internet of Things (IoT).

5.2.1 Integration Technology: ChipFilm Patch as Interposer

Several challenges and trade-offs are tackled for the successful integration of the smart label. Chip foundries use different technology stacks, metal layers, passivation material, pad size, and pitch, so all these factors greatly influence the chip thinning and embedding. Figures 5.2b and 5.2c show two photographs of a CFP wafer in which the foil is still attached to the carrier wafer and two 30 μm-thick chips embedded using CFP. Here, thin chips embedding in CFP can follow either face-up or face-down approaches.[88], [14] The advantage of the face-up approach is that the CFP can be processed directly on the carrier wafer from the start until the final release of the foil. However, differences in chip thickness, in case of embedding two or more chips in the same package, result in excessive surface topography, thus requiring thicker embedding polymer, which also limits the interconnect metal pitch to about 10 μm (Figures 5.2d and 5.2e). In the face-down approach, the chips are temporarily assembled on a first carrier wafer. In this way, the chip pad layers are leveled, which allows for better surface topography and micrometer-pitch global interconnects. After attaching a second carrier wafer substrate, that first substrate is removed and the CFP process continues in the same steps as the face-up approach.

Due to the complexity of the targeted HySiF and the economics of embedding more than two chips in the same foil, the CFP is used here as a fan-out package for thin chip(s) instead of using it for realizing the whole sensor system, as previously demonstrated in Section 5.1. CFP using the face-up approach is chosen for this demonstrator. To minimize the uncertainty in the

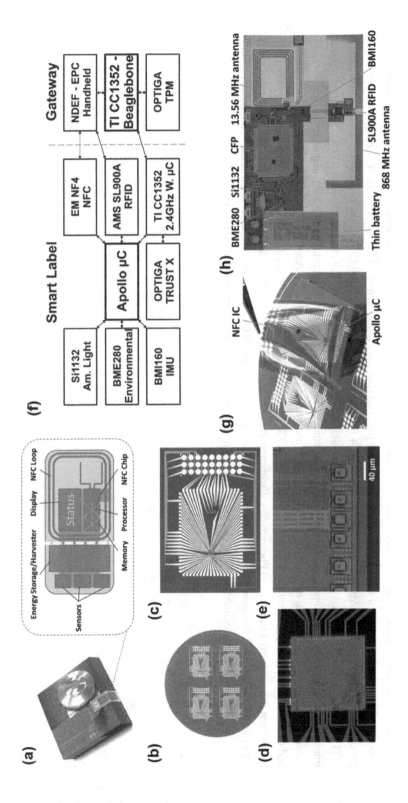

Caption for Figure 5.2 (cont.)

Figure 5.2(a) Conceptual illustration of a smart label based on HySiF for logistics tracking. This smart label is developed within the ParsiFAL4.0 project, in which the company Bosch GmbH is a partner.[120] (b) Photograph of a CFP wafer after finalizing all its fabrication process. (c) Photograph of the Apollo microcontroller and NF4 RFID tag chips embedded using CFP. (d) Micrograph of a Texas Instruments (TI) microcontroller embedded in CFP. (e) Fine-pitch fan-out CFP after via opening using dry etching process. (f) Block diagram of the implemented smart label developed within the ParsiFAL4.0 project. Temperature, humidity, pressure (BME280), and ambient light (Si1132) sensors in addition to an inertial measurement unit (BMI160) are integrated on the smart label. Apollo μC is used as a central data collection and distribution hub, which connects to a wireless μC (TI CC1352) for the 2.4 GHz 6LowPAN connectivity, which is secured by OPTIGA chips from the company Infineon. (g) Photography illustrating the simple release of the CFP with the thin Apollo and NFC embedded chips. (h) Flexible smart label for activity monitoring using LCP as main substrate and CFP as interposer, assembled using foam adhesive and printed interconnects. Other electronic components are either integrated into or soldered on the LCP substrate.

final chip thickness, the two bare die Apollo and NF4 chips are thinned together.

5.2.2 System Design

Figure 5.2f shows the block diagram of the high-performance smart label system developed within the ParsiFAL4.0 project.[120] Apollo μC is a low-power microcontroller usually integrated into wearable electronics. Here, it is the main μC acting as a central data collection and distribution node by using conventional wired communication such as SPI and I2 C. An integrated temperature, humidity, and pressure sensor (BME280) is used to monitor environmental parameters and connects to the Apollo μC using I2 C. An inertial measurement unit (BMI160) is used to detect sudden movements and collisions and connects to the Apollo μC via SPI. An ambient light sensor (Si1132) is integrated and connects to the Apollo μC using I2 C.

The sensory data are organized according to the IoT-compatible IPSO data model and transmitted to a central gateway using the TI CC1352 6LowPAN 2.4 GHz wireless microcontroller. For secure communication and authentication, a product from the company Infineon Technologies AG called OPTIGA TRUST is used.[152] It ensures secure communication and mutual authentication with cryptographic technology between the smart label and gateway. The NF4 chip, manufactured by the company EM Microelectronic,[153] implements the NFC protocols required for secure near-field communication. It is used for configuring and debugging the smart labels and allows for fast label identification. A longer-range 868 MHz EPC-compliant RFID tag from the company AMS is also used for monitoring the label status and reporting a summary of the labels' activities when needed.[154]

5.2.3 Demonstrator

Figure 5.2g shows the simple release of the CFP from the carrier wafer including the thin Apollo μC and NF4 NFC embedded chips. Figure 5.2h shows the flexible smart label in which the CFP is then used as an interposer that is assembled using foam adhesive on liquid crystal polymer (LCP) substrate manufactured by the company Würth Elektronik GmbH. Printed interconnects are then used to connect the CFP with the LCP motherboard, which includes other sensors, μC ICs, and communication chips. A 3 V thin battery is used to operate the smart label and is soldered to the LCP motherboard. 13.56 MHz and 868 MHz antennas are integrated into the LCP motherboard. Although a fully flexible and bendable label is not completely achieved, this high-performance demonstrator represents a unique solution featuring innovation in flexible thin-chip packaging, integration technology, and system design fields.

Abbreviations

ACA	Anisotropic Conductive Adhesive
ADC	Analog-to-Digital Converter
BCB	Benzocyclobutene
BGA	Ball-Grid Array
CFP	ChipFilm Patch
CNT	Carbon Nanotube
CTAT	Complementary to Absolute Temperature
DAC	Digital-to-Analog Converter
DBG	Dicing-before-Grinding
DDA	Differential Difference Amplifier
DNL	Differential Nonlinearity
ECG	Electrocardiogram
ECT	Embedded Component Technology
ESD	Electrostatic Discharge
EVB	Evaluation Board
E-Skin	Electronic Skin
FoM	Figure-of-Merit
GPIO	General-Purpose Input/Output
HFRC	High-Frequency RC Oscillator
HySiF	Hybrid System-in-Foil
IC	Integrated Circuit
InGaZno	Indium-Gallium-Zinc-Oxide
INL	Integral Nonlinearity
IoT	Internet of Things
LCP	Liquid Crystal Polymer
LED	Light Emitting Diodes
LFRC	Low-Frequency RC Oscillator
LSB	Least Significant Bit
μC	Microcontroller
MCM	Multichip Module
MEMS	Micro-Electromechanical Systems
MIM	Metal Insulator Metal
MM	More Moore
MOS	Metal Oxide Semiconductor
MtM	More than Moore
NFC	Near-Field Communication

OTFT	Organic Thin-Film Transistors
PCB	Printed Circuit Board
PDMS	Polydimethylsiloxane
PEN	Polyethylene Naphthalate
PI	Polyimide
PTAT	Proportional to Absolute Temperature
PVDF	Polyvinylidene Fluoride
PVT	Process, Supply Voltage, Temperature
RFID	Radio-Frequency Identification
RH	Relative Humidity
RTD	Resistance Temperature Detector
R2R	Roll-to-Roll
SAR	Successive-Approximation Register
SiP	System-in-Package
SMU	Source Measurement Unit
SOI	Silicon on Insulator
SoC	System-on-Chip
SoP	System-on-Package
SWO	Single-Wire Output
TFT	Thin-Film Transistors
UTC	Ultrathin Chip
UTCP	Ultrathin Chip Package
VLSI	Very-Large-Scale Integration

References

[1] G. E. Moore, "Cramming more components onto integrated circuits," *Electronics*, vol. 38, no. 8, pp. 114 ff., April 1965.

[2] P. Wong, keynote speech at the Hot Chips conference, Stanford University, August 18–20, 2019.

[3] S. K. Moore, "Another step toward the end of Moore's law: Samsung and TSMC move to 5-nanometer manufacturing," *IEEE Spectrum*, vol. 56, no. 6, pp. 9–10, May 2019.

[4] P. Ye, T. Ernst, and M. V. Khare, "The last silicon transistor: Nanosheet devices could be the final evolutionary step for Moore's Law," *IEEE Spectrum*, vol. 56, no. 8, pp. 30–35, August 2019.

[5] W. Arden et al., "More-than-Moore White Paper," *White Paper in International Technology Roadmap for Semiconductors (ITRS)*, 2010. www.yumpu.com/s/X5NAyREt1px0eEfN

[6] Y. Afsar et al., "15.1 Large-scale acquisition of large-area sensors using an array of frequency-hopping ZnO thin-film-transistor oscillators," *Proceedings of the IEEE International Solid-State Circuits Conference* (ISSCC), pp. 256–257, February 2017.

[7] T. Moy et al., "An EEG acquisition and biomarker-extraction system using low-noise-amplifier and compressive-sensing circuits based on flexible, thin-film electronics," *IEEE Journal of Solid-State Circuits*, vol. 52, no. 1, pp. 309–321, January 2017.

[8] A. V. Quintero et al., "Smart RFID label with a printed multisensor platform for environmental monitoring," *Flexible Printed Electronics*, vol. 1, no. 2, pp. 1–13, January 6, 2016.

[9] N. P. Papadopoulos et al., "Toward temperature tracking with unipolar metal-oxide thin-film SAR C-2 C ADC on plastic," *IEEE Journal of Solid-State Circuits*, vol. 53, no. 8, pp. 2263–2272, August 2018.

[10] Z. Ma et al., "Materials and design considerations for fast flexible and stretchable electronics," *Proceedings of the IEEE International Electron Devices Meeting (IEDM)*, Washington, DC, February 16, 2015, pp. 19.2.1–19.2.4.

[11] Y. H. Jung, H. Zhang, S. J. Cho, and Z. Ma, "Flexible and stretchable microwave microelectronic devices and circuits," *IEEE Transactions on Electron Devices*, vol. 64, no. 5, pp. 1881–1893, May 2017.

[12] C. Harendt et al., "Hybrid Systems in Foil (HySiF) exploiting ultra-thin flexible chips," *44th European Solid State Device Research Conference (ESSDERC)*, Venice, 2014, pp. 210–213.

[13] J. N. Burghartz et al., "Hybrid Systems-in-Foil – combining the merits of thin chips and of large-area electronics," *IEEE Journal of the Electron Devices Society*, vol. 7, pp. 776–783, 2019.

[14] J. N. Burghartz et al., "Hybrid Systems-in-Foil – combining the merits of thin chips and of large-area electronics," *IEEE Journal of the Electron Devices Society*, vol. 7, pp. 776–783, 2019.

[15] M. Jung et al., "Amorphous FeZr metal for multi-functional sensor in electronic skin," *npj Flexible Electronics*, vol. 3, April 2019, 8.

[16] D. Sun et al., "Flexible high-performance carbon nanotube integrated circuits," *Nature Nanotechnology*, vol. 6, pp. 156–161, February 2011.

[17] Y. Sun, M. et al., "Flexible organic photovoltaics based on water-processed silver nanowire electrodes," *Nature Electronics*, 2, pp. 513–520, November 2019.

[18] W. Li, S. Yang, and A. Shamim, "Screen printing of silver nanowires: Balancing conductivity with transparency while maintaining flexibility and stretchability," *npj Flexible Electronics*, vol. 3, no. 13, December 2019. DOI: https://doi.org/10.1038/s41528-019-0057-1

[19] J. N. Burghartz et al., "Hybrid Systems-in-Foil – combining the merits of thin chips and of large-area electronics," *IEEE Journal of the Electron Devices Society*, vol. 7, pp. 776–783, 2019.

[20] K. Myny et al., "An 8b organic microprocessor on plastic foil," *IEEE International Solid-State Circuits Conference*, San Francisco, 2011, pp. 322–324.

[21] K. Myny et al., "15.2 A flexible ISO14443-A compliant 7.5 mW 128b metal-oxide NFC barcode tag with direct clock division circuit from 13.56 MHz carrier," *Proceedings of the IEEE International Solid-State Circuits Conference (ISSCC)*, San Francisco, 2017, pp. 258–259.

[22] M. Elsobky et al., "Ultra-thin smart electronic skin based on hybrid system-in-foil concept combining three flexible electronics technologies," *Electronics Letters*, vol. 54, no. 6, pp. 338–340, March 22, 2018.

[23] M. Elsobky et al., "Characterization of thin-film temperature sensors and ultra-thin chips for HySiF integration," *2019 IEEE International Conference on Flexible and Printable Sensors and Systems (FLEPS)*. DOI: https://doi.org/10.1109/FLEPS.2019.8792313.

[24] Y. Mahsereci, S. Sailer, H. Richter, and J. Burghartz, "16.1 An ultra-thin flexible CMOS stress sensor demonstrated on an adaptive robotic gripper," *Proceedings of the IEEE International Solid-State Circuits Conference – (ISSCC) Digest of Technical Papers*, San Francisco, February 22–26, 2015, pp. 1–3.

[25] M. Elsobky et al., "Ultra-thin smart electronic skin based on hybrid system-in-foil concept combining three flexible electronics technologies," *Electronics Letters*, vol. 54, no. 6, pp. 338–340, March 22, 2018.

[26] M. Elsobky et al., "Ultra-thin sensor systems integrating silicon chips with on-foil passive and active components," *Proceedings*, vol. 2, p. 748, December 2018.

[27] B. S. Cook, J. R. Cooper, and M. M. Tentzeris, "Multi-Layer RF capacitors on flexible substrates utilizing inkjet printed dielectric polymers," *IEEE Microwave and Wireless Components Letters*, vol. 23, no. 7, pp. 353–355, July 2013.

[28] T. Widlund, S. Yang, Y.-Y. Hsu, and N. Lu, "Stretchability and compliance of freestanding serpentine-shaped ribbons," *International Journal of Solids and Structures*, vol. 51, August 2014.

[29] H. Hocheng and C.-M. Chen, "Design, fabrication and failure analysis of stretchable electrical routings," *Sensors*, vol. 14, pp. 11855–11877, July 2014.

[30] T. T. Nguyen et al., "A flexible bimodal sensor arrays for simultaneous sensing of pressure and temperature," *Advanced Materials*, vol. 26, no. 5, pp. 796–804, February 2014.

[31] Q. Hua et al., "Skin-inspired highly stretchable and conformable matrix networks for multifunctional sensing," *Nature Communications*, vol. 9, December 2018.

[32] K. Kim et al., "Low-voltage, high-sensitivity and high-reliability bimodal sensor array with fully inkjet-printed flexible conducting electrode for low power consumption electronic skin," *Nano Energy*, vol. 41, pp. 301–307, November 2017.

[33] P. Escobedo, M. Bhattacharjee, F. Nikbakhtnasrabadi, and R. Dahiya, "Smart bandage with wireless strain and temperature sensors and battery-less NFC tag," *IEEE Internet of Things Journal*, vol. 8, no.6, pp. 5093–5100, March 15, 2021.

[34] H. Yang et al., "Wireless Ti3C2Tx MXene strain sensor with ultrahigh sensitivity and designated working windows for soft exoskeletons," *ACS Nano*, vol. 14, no. 9, pp. 11860–11875, August 11, 2020.

[35] B. Tian et al., "Three-dimensional flexible nanoscale field-effect transistors as localized bioprobes," *Science*, vol. 329, pp. 830–834, August 2010.

[36] D. Khodagholy et al., "In vivo recordings of brain activity using organic transistors," *Nature Communications*, vol. 4, no. 1575, March 12, 2013. DOI: https://doi.org/10.1038/ncomms2573

[37] J. T. Robinson et al., "Developing next-generation brain sensing technologies – a review," *IEEE Sensors Journal*, vol. 19, no. 22, 2019. DOI: https://doi.org/10.1109/jsen.2019.2931159.

[38] L. Yu et al., "High-yield large area MoS2 technology: Material, device and circuits co-optimization," *2016 IEEE International Electron Devices Meeting (IEDM)*, San Francisco, 2016, pp. 5.7.1–5.7.4.

[39] S. Ahmed, F. A. Tahir, A. Shamim, and H. M. Cheema, "A compact Kapton-based inkjet-printed multiband antenna for flexible wireless devices," *IEEE Antennas and Wireless Propagation Letters*, vol. 14, pp. 1802–1805, 2015.

[40] M. Tang, T. Shi, and R. W. Ziolkowski, "Flexible efficient quasi-yagi printed uniplanar antenna," *IEEE Transactions on Antennas and Propagation*, vol. 63, no. 12, pp. 5343–5350, December 2015.

[41] C. Yi-Bing, P. Alex, H. Fuzhi, and P. Yong, "Print flexible solar cells," *Springer Nature*, vol. 539, no. 7630, pp. 488–489, November 2016.

[42] M. Ntagios, P. Escobedo, and R. Dahiya, "3D printed packaging of photovoltaic cells for energy autonomous embedded sensors," *2020 IEEE SENSORS*, Rotterdam, Netherlands, 2020, pp. 1–4.

[43] C. Garcia Nunez, L. Manjakkal, and R. Dahiya, "Energy autonomous electronic skin," *npj Flexible Electronics*, vol. 3, December 2019. DOI: https://doi.org/10.1038/s41528-018-0045-x

[44] Y. Hu and X. Sun, "Flexible rechargeable lithium ion batteries: Advances and challenges in materials and process technologies," *J. Mater. Chem. A*, vol. 2, pp. 10712–10738, July 28, 2014.

[45] H. Cha, J. Kim, Y. Lee, J. Cho, and M. Park, "Issues and challenges facing flexible lithium-ion batteries for practical application," *Small*, vol. 14, December 2017, e170 2989. DOI: https://doi.org/10.1002/smll.201702989

[46] L. Manjakkal, A. Pullanchiyodan, N. Yogeswaran, E. S. Hosseini, and R. Dahiya, "A wearable supercapacitor based on conductive PEDOT:PSS-coated cloth and a sweat electrolyte,," *Adv. Mater.* 2020, vol. 32, no. 24, 1907254. DOI: https://doi.org/10.1002/adma.201907254

[47] E. Ozer et al., "Bespoke machine learning processor development framework on flexible substrates," *2019 IEEE International Conference on Flexible and Printable Sensors and Systems (FLEPS)*, Glasgow, UK, July 8–10, 2019, pp. 1–3.

[48] White Paper, www.xilinx.com/support/documentation/white_papers/wp505-versal-acap.pdf.

[49] Intel Newsroom, July 10, 2019, https://newsroom.intel.de/news/intel-unveils-new-tools-in-its-advanced-chip-packaging-toolbox/?wapkw=fo veros#gs.99bfyu.

[50] White Paper, www.achronix.com/wp-content/uploads/2019/03/Chiplets-Taking-SoC-Design-Where-no-Monolithic-IC-has-Gone-Before-WP016. pdf.

[51] S. Gupta, W. T. Navaraj, L. Lorenzelli, and R. Dahiya, "Ultra-thin chips for high-performance flexible electronics," *npj Flexible Electronics*, vol. 2, no. 8, March 14, 2018. DOI: https://doi.org/10.1038/s41528-018-0021-5.

[52] P. Ruther, S. Herwik, S. Kisban, K. Seidl, and O. Paul, "Recent progress in neural probes using silicon MEMS technology," *IEEJ Transactions on Electrical and Electronic Engineering*, vol. 5, no. 5, pp. 505–515, 2010.

[53] G. A. T. Sevilla, S. B. Inayat, J. P. Rojas, A. M. Hussain, and M. M. Hussain, "Flexible and semi-transparent thermoelectric energy harvesters from low cost bulk silicon (100)}," *Small*, vol. 9, no. 23, pp. 3916–3921, 2013.

[54] A. Schander et al., "Design and fabrication of novel multi-channel floating neural probes for intracortical chronic recording," *Sensors and Actuators A: Physical*, vol. 247, pp. 125–135, August 15, 2016.

[55] NXP Semiconductors, "ePassports – secure chip based travel documents," www.nxp.com/applications/solutions.

[56] S. Takyu, T. Kurosawa, N. Shimizu, and S. Harada, "Novel wafer dicing and chip thinning technologies realizing high chip strength," *Proceedings of the 56th Electronic Components and Technology Conference 2006*, San Diego, 2006. DOI: https://doi.org/10.1109/ECTC.2006.1645874.

[57] M. Hassan, E. A. Angelopoulos, H. Rempp, S. Endler, and J. N. Burghartz, "Packaging challenges associated with warpage of ultra-thin chips," *Proceedings of the 3rd Electronics System Integration Technology Conference ESTC*, Berlin, pp. 505–515, 2010. DOI: https://doi.org/10 .1109/ESTC.2010.5642819

[58] W. R. Morcom et al. Self-supported ultra thin silicon wafer process. US patent 6162702 A (2000).

[59] M. K. Weldon et al., "Mechanism of silicon exfoliation by hydrogen implantation and He, Li and Si co-implantation [SOI technology]," 1997 *IEEE International SOI Conference Proceedings*, Fish Camp, CA, 1997, pp. 124–125.

[60] C. Braley, "Si exfoliation by MeV proton implantation," *Nuclear Instruments and Methods in Physics Research Section B: Beam Interactions with Materials and Atoms*, vol. 277, pp. 93–97, April 2012.

[61] D. Shahrjerdi and S. W. Bedell, "Extremely flexible nanoscale ultrathin body silicon integrated circuits on plastic," *Nano Letters*, vol. 13, no. 1, pp. 315–320, January 2013.

[62] J. N. Burghartz, "You can't be too thin or too flexible," *IEEE Spectrum*, vol. 50, no. 3, pp. 38–61, March 2013.

[63] Y. Kato et al., "A large-area, flexible, ultrasonic imaging system with a printed organic transistor active matrix," *Proceedings of the IEEE International Electron Devices Meeting*, San Francisco. DOI: https://doi .org/10.1109/TED.2010.2044278.

[64] M. Ahmed, D. P. Butler, and Z. Celik-Butler, "MEMS absolute pressure sensor on a flexible substrate," *Proceedings of the IEEE 25th International Conference on Micro Electro Mechanical Systems (MEMS)*, Paris, 2012, pp. 575–578.

[65] B. Gao, M. M. F. Yuen, and T. T. Ye, "Flexible frequency selective metamaterials for microwave applications," *Scientific Reports*, vol. 7, March 2017. DOI: https://doi.org/10.1038/srep45108

[66] R. Augustine et al., "SAR reduction of wearable antennas using polymeric ferrite sheets," *Proceedings of the Fourth European Conference on Antennas and Propagation*, Barcelona, April 12–16, 2010, pp. 1–3.

[67] S. Ahmed, F. A. Tahir, A. Shamim, and H. M. Cheema, "A compact Kapton-based inkjet-printed multiband antenna for flexible wireless devices," *IEEE Antennas and Wireless Propagation Letters*, vol. 14, pp. 1802–1805, 2015.

[68] M. Tang, T. Shi, and R. W. Ziolkowski, "Flexible efficient quasi-yagi printed uniplanar antenna," *IEEE Transactions on Antennas and Propagation*, vol. 63, no. 12, pp. 5343–5350, December 2015.

[69] A. Kiourti and J. L. Volakis, "Stretchable and flexible e-fiber wire antennas embedded in polymer," *IEEE Antennas and Wireless Propagation Letters*, vol. 13, pp. 1381–1384, 2014.

[70] R. B. V. B. Simorangkir, Y. Yang, K. P. Esselle, and B. A. Zeb, "A method to realize robust flexible electronically tunable antennas using polymer-embedded conductive fabric," *IEEE Transactions on Antennas and Propagation*, vol. 66, no. 1, pp. 50–58, January 2018.

[71] Su-Tsai Lu and Wen-Hwa Chen, "Reliability of ultra-thin chip-on-flex (UTCOF) with anisotropic conductive adhesive (ACA) joints," *Proceedings of the 58th Electronic Components and Technology Conference,* Lake Buena Vista, FL, 2008, pp. 1287–1293.

[72] C. Banda, R. W. Johnson, T. Zhang, Z. Hou, and H. K. Charles, "Flip chip assembly of thinned silicon die on flex substrates," *IEEE Transactions on Electronics Packaging Manufacturing*, vol. 31, no. 1, pp. 1–8, January 2008.

[73] V. Marinov et al., "Laser-enabled advanced packaging of ultrathin bare dice in flexible substrates," *IEEE Transactions on Components,*

Packaging and Manufacturing Technology, vol. 2, no. 4, pp. 569–577, April 2012.

[74] J. Wolf et al., "Ultra-thin silicon chips in flexible microsystems," *Proceedings of the 13th Electronic Circuits World Convention ECWC,* Nuremberg, Germany, May 2014.

[75] A. Sridhar, M. Cauwe, H. Fledderus, R. H. L. Kusters, and J. van den Brand, "Novel interconnect methodologies for ultra-thin chips on foils," *2012 IEEE 62nd Electronic Components and Technology Conference*, San Diego, 2012, pp. 238–244.

[76] J. van den Brand, J. de Baets, T. van Mol, and A. Dietzel, "Systems-in-Foil – Devices, fabrication processes and reliability issues," *Journal of Microelectronics Reliability*, vol. 48, no. 8–9, pp. 1123–1128, 2008.

[77] J. Govaerts, W. Christiaens, E. Bosman, and J. Vanfleteren, "Fabrication processes for embedding thin chips in flat flexible substrates," *IEEE Transactions on Advanced Packaging*, vol. 32, no. 1, pp. 77–83, February 2009.

[78] Tzu-Ying Kuo et al., "Flexible and ultra-thin embedded chip package," *2009 59th Electronic Components and Technology Conference,* San Diego, 2009, pp. 1749–1753.

[79] W. Christiaens, E. Bosman, and J. Vanfleteren, "UTCP: A novel polyimide-based ultra-thin chip packaging technology," *IEEE Transactions on Components and Packaging Technologies*, vol. 33, no. 4, pp. 754–760, December 2010.

[80] N. Palavesam, W. Hell, A. Drost, C. Landesberger, C. Kutter, and K. Bock, "A novel low cost roll-to-roll manufacturing compatible ultra-thin chip integration and direct metal interconnection process for flexible hybrid electronics," *2019 IMAPS Nordic Conference on Microelectronics Packaging (NordPac)*, pp. 6–11, 2019. doi: https://doi.org/10.23919/NORDPAC.2019.8760350.

[81] L. Boettcher, D. Manessis, A. Ostmann, S. Karaszkiewicz, and H. Reichl, "Embedding of chips for System in Package realization – technology and applications," *2008 3rd International Microsystems, Packaging, Assembly & Circuits Technology Conference*, Taipei, Taiwan, pp. 383–386, October 22–24, 2008.

[82] D. Manessis, L. Boettcher, S. Karaszkiewicz, A. Ostmann, R. Aschenbrenner, and K. Lang, "Chip embedding technology developments leading to the emergence of miniaturized system-in-packages," *Proceedings of the 18th European Microelectronics & Packaging Conference*, Brighton, UK, pp. 1–8, September 12–15, 2011.

[83] T. Sterken et al., "Ultra-Thin Chip Package (UTCP) and stretchable circuit technologies for wearable ECG system," *2011 Annual International Conference of the IEEE Engineering in Medicine and Biology Society*, 2011, pp. 6886-6889, doi: https://doi.org/10.1109 /IEMBS.2011.6091734.

[84] J. N. Burghartz and C. Harendt, "Method for producing an integrated circuit and resulting film chip," US Patent 20120161293 A1, issue date: July 6, 2012.

[85] M. Elsobky et al., "Characterization of on-foil sensors and ultra-thin chips for HySiF integration," in *IEEE Sensors Journal*, vol. 20, no. 14, pp. 7595–7604, July 15, 2020.

[86] M. Elsobky et al., "Ultra-thin smart electronic skin based on hybrid system-in-foil concept combining three flexible electronics technologies," *Electronics Letters*, vol. 54, no. 6, pp. 338–340, March 2018.

[87] M. Hassan et al., "Combining organic and printed electronics in Hybrid System in Foil (HySiF) based smart skin for robotic applications," *Proceedings of the 2015 European Microelectronics Packaging Conference (EMPC)*, Friedrichshafen, Germany, pp. 1–6, September 14–16, 2015.

[88] M. Hassan, C. Schomburg, C. Harendt, E. Penteker, and J. N. Burghartz, "Assembly and embedding of ultra-thin chips in polymers," *Proceedings of the 2013 European Microelectronics Packaging Conference (EMPC)*, Grenoble, France. pp. 1–6, September 9–12, 2013.

[89] G. Alavi, H. Sailer, B. Albrecht, C. Harendt, and J. N. Burghartz, "Adaptive layout technique for microhybrid integration of chip-film patch," *IEEE Transactions on Components, Packaging and Manufacturing Technology*, vol. 8, no. 5, pp. 802–810, May 2018.

[90] N. P. Papadopoulos et al., "Toward temperature tracking with unipolar metal-oxide thin-film SAR C-2 C ADC on plastic," *IEEE Journal of Solid-State Circuits*, vol. 53, no. 8, pp. 2263–2272, August 2018.

[91] X. Leng, W. Li, D. Luo and F. Wang, "Differential structure with graphene oxide for both humidity and temperature sensing," *IEEE Sensors Journal*, vol. 17, no. 14, pp. 4357–4364, 2017.

[92] Webb, R. Chad et al., "Ultrathin conformal devices for precise and continuous thermal characterization of human skin," *Nature Materials*, vol. 12, pp. 938–944, 2013.

[93] S. Pan and K. A. A. Makinwa, "A 0.25 mm^2-resistor-based temperature sensor with an inaccuracy of 0.12 °C (3σ) from −55 °C to 125 °C," *IEEE Journal of Solid-State Circuits*, vol. 53, no. 12, pp. 3347–3355, December 2018.

[94] M. Soni, M. Bhattacharjee, M. Ntagios, and R. Dahiya, "Printed temperature sensor based on PEDOT: PSS-graphene oxide composite," *IEEE
Sensors Journal*, vol. 20, no. 14, pp. 7525–7531, July 15, 2020.

[95] M. Elsobky et al. , "Characterization of thin-film temperature sensors and
ultra-thin chips for HySiF integration," *Proceedings of IEEE International
Conference on Flexible and Printable Sensors and Systems (FLEPS)*,
pp. 1–3, 2019. doi: https://doi.org/10.1109/FLEPS.2019.8792313.

[96] A. Modafe, N. Ghalichechian, M. Powers, M. Khbeis, and R. Ghodssi,
"Embedded benzocyclobutene in silicon: An integrated fabrication pro-
cess for electrical and thermal isolation in MEMS," *Microelectronic
Engineering*, vol. 82, pp. 157–167, 2005.

[97] I. J. Malik et al., "Surface roughness of silicon wafers on different lateral
length scales," *Journal Electrochemical Society*, vol. 140, no. 5, pp.
75–77, May 1993.

[98] K. Szendrei et al., "Touchless optical finger motion tracking based on 2D
nanosheets with giant moisture responsiveness," *Advanced Materials*, vol.
27, pp. 6341–6348, 2015.

[99] M. Elsobky et al., "Ultra-thin relative humidity sensors for hybrid
system-in-foil applications," *Proceedings of IEEE SENSORS*,
Glasgow, 2017.

[100] M. Elsobky et al., "Relative humidity sensors for system-in-foil applications," *Proceedings of MikroSystemTechnik Congress*, Munich, Germany,
2017.

[101] Y. Mahsereci, S. Saller, H. Richter, and J. N. Burghartz, "An ultra-thin
flexible CMOS stress sensor demonstrated on an adaptive robotic
gripper," *IEEE Journal of Solid-State Circuits*, vol. 51, no. 1, pp.
273–280, January 2016.

[102] M. Kuhl et al., "A wireless stress mapping system for orthodontic
brackets using CMOS integrated sensors," *IEEE Journal of Solid-State
Circuits*, vol. 48, no. 9, pp. 2191–2202, September 2013.

[103] S. Huber, W. Leten, M. Ackermann, C. Schott, and O. Paul, "A fully
integrated analog compensation for the piezo-Hall effect in a CMOS
single-chip Hall sensor microsystem," *IEEE Sensors Journal*, vol. 15,
no. 5, pp. 2924–2933, May 2015.

[104] C. Pang et al., "A flexible and highly sensitive strain-gauge sensor using
reversible interlocking of nanofibers," *Nature Materials*, vol. 11, pp.
795–801, September 2012.

[105] F. Yin et. al., "Highly sensitive and transparent strain sensors with an
ordered array structure of AgNWs for wearable motion and health
monitoring," *Scientific Reports*, vol. 9, February 2019.

[106] H. Guo et al., "Vectorial strain gauge method using single flexible orthogonal polydimethylsiloxane gratings," *Scientific Reports*, vol. 6, March 2013.

[107] M. Elsobky, Y. Mahsereci, J. Keck, H. Richter, and J. N. Burghartz, "Design of a CMOS readout circuit on ultra-thin flexible silicon chip for printed strain gauges," *Advances in Radio Science*, vol. 15, pp. 123–130, September 2017.

[108] Y.-H. Wen et al., "Mechanically robust micro-fabircated strain gauges for use on bones," *Proceedings of the Conference on Microtechnologies in Medicine and Biology*, pp. 302–304, 2005.

[109] R. J. Stephen, K. Rajanna, V. Dhar, K. K. Kumar, and S. Nagabushanam, "Thinfilm strain gauge sensors for ion thrust measurement," *IEEE Sensors Journal*, vol. 4, pp. 373–377, 2004.

[110] J. A. Paulsen and M. J. Renn, *Maskless Printing of Miniature Polymer Thick Film Resistors for Embedded Applications*. Optomec, Inc., 2014.

[111] M. Hedges, M. Renn, and M. Kardos, "Mesoscale deposition technology for electronics applications," *Proceedings of the 5th International Conference on Polymers and Adhesives*, pp. 53–57, 2005. doi: https://doi .org/10.1109/POLYTR.2005.1596486.

[112] B. Ando, S. Baglio, S. L. Malfa, and G. L'Episcopo, "All inkjet printed system for strain measurement," *IEEE Sensors Journal*, vol. 13, pp. 4874–4879, 2013.

[113] A. Kiourti and J. L. Volakis, "Stretchable and flexible e-fiber wire antennas embedded in polymer," *IEEE Antennas and Wireless Propagation Letters*, vol. 13, pp. 1381–1384, 2014.

[114] T. Meister et al., "Program FFlexCom – High frequency flexible bendable electronics for wireless communication systems," *Proceedings of the IEEE International Conference on Microwaves, Antennas, Communications and Electronic Systems (COMCAS)*, pp. 1–6, 2017. doi: https://doi.org/10.1109/COMCAS.2017.8244733.

[115] M. Elsobky et al., "Ultra-thin sensor systems integrating silicon chips with on-foil passive and active component," *Proceedings*, vol. 2, no. 748, December 2018.

[116] G. Di Massa, S. Costanzo, A. Borgia, F. Venneri, and I. Venneri, "Innovative dielectric materials at millimeter-frequencies," *Proceedings of the 20th International Conference on Applied Electromagnetics and Communications (ICECom)*, Dubrovnik, 2010.

[117] G. Alavi, "Hybrid system-in-foil integration and interconnection technology based on adaptive layout technique," University of Stuttgart, Ph.D. dissertation, 2019.

[118] G. Alavi et al., "Embedding and interconnecting of ultra-thin rf chip in combination with flexible wireless hub in polymer foil," *Proceedings of the 7th Electronic* System-Integration Technology Conference (ESTC), pp. 1–5, 2018, .

[119] S. Ha, C. Kim, J. Park, S. Joshi, and G. Cauwenberghs, "Energy recycling telemetry IC with simultaneous 11.5 mW power and 6.78 Mb/s backward data delivery over a single 13.56 MHz inductive link," *IEEE Journal of Solid-State Circuits*, vol. 51, no. 11, pp. 2664–2678, November 2016. doi: https://doi.org/10.1109/JSSC.2016.2600864.

[120] Produktfähige autarke und sichere Foliensysteme für Automatisierungslösungen in Industrie 4.0, Project Website [online], www.parsifal40.de, accessed March 5, 2021.

[121] G. Qi, P. Mak, and R. P. Martins, "26.9 A 0.038 mm2 SAW-less multi-band transceiver using an N-Path SC gain loop," *Proceedings of the IEEE International Solid-State Circuits Conference (ISSCC)*, San Francisco, pp. 452–454, 2016.

[122] Y. Tseng and T. Ma, "On-chip GIPD bandpass filter using synthesized stepped impedance resonators," *IEEE Microwave and Wireless Components Letters*, vol. 24, no. 3, pp. 140–142, March 2014.

[123] S. Bandyopadhyay, P. P. Mercier, A. C. Lysaght, K. M. Stankovic, and A. P. Chandrakasan, "A 1.1 nW energy-harvesting system with 544 pW quiescent power for next-generation implants," *IEEE Journal of Solid-State Circuits*, vol. 49, no. 12, pp. 2812–2824, December 2014.

[124] S. Ha, C. Kim, J. Park, S. Joshi, and G. Cauwenberghs, "Energy recycling telemetry IC with simultaneous 11.5 mW power and 6.78 Mb/s backward data delivery over a single 13.56 MHz inductive link," *IEEE Journal of Solid-State Circuits*, vol. 51, no. 11, pp. 2664–2678, November 2016. doi: https://doi.org/10.1109/JSSC.2016.2600864.

[125] S. M. Won et al., "Piezoresistive strain sensors and multiplexed arrays using assemblies of single-crystalline silicon nanoribbons on plastic substrates," *IEEE Transactions on Electron Devices*, vol. 58, no. 11, pp. 4074–4078, November 2011.

[126] M. Elsobky et al., "A digital library for a flexible low-voltage organic thin-film transistor technology," *Organic Electronics*, vol. 50, pp. 491–498, November 2017.

[127] T. Ghani et al., "A 90 nm high volume manufacturing logic technology featuring novel 45 nm gate length strained silicon CMOS transistors," *Proceedings of the IEEE International Electron Devices Meeting*, Washington, DC, 2003.

[128] H. Heidari, N. Wacker, and R. Dahiya, "Bending induced electrical response variations in ultra-thin flexible chips and device modeling," *Applied Physics Reviews*, vol. 4, 2017.

[129] M. Hassan et al., "Anomalous stress effects in ultra-thin silicon chips on foil," *Proceedings of the IEEE International Electron Devices Meeting (IEDM)*, Baltimore, December 7–9, 2009.

[130] Stoney, G. G., 1909, "The tension of metallic films deposited by electrolysis," *Proc. R. Soc. London, Ser. A*, 82, pp. 172–175.

[131] T. Ishihara, K. Suzuki, S. Suwazono, M. Hirata, and H. Tanigawa, "CMOS integrated silicon pressure sensor," *IEEE Journal of Solid-State Circuits*, vol. 22, no. 2, pp. 151–156, April 1987.

[132] E. Yoon and K. D. Wise, "An integrated mass flow sensor with on-chip CMOS interface circuitry," *IEEE Transactions on Electron Devices*, vol. 39, no. 6, pp. 1376–1386, June 1992.

[133] J. N. Burghartz, *Ultra-thin Chip Technology and Applications*. Springer, 2011.

[134] S. Gupta, W. T. Navaraj, L. Lorenzelli, and R. Dahiya, " Ultra-thin chips for high-performance flexible electronics," *npj Flexible Electronics*, vol. 2, no. 1, March 2018.

[135] S. Pan, Y. Luo, S. Heidary Shalmany, and K. A. A. Makinwa, "A resistor-based temperature sensor with a 0.13 pJ · K2 resolution FoM," *IEEE Journal of Solid-State Circuits*, vol. 53, no. 1, pp. 164–173, January 2018.

[136] P. Harpe, "A compact 10-b SAR ADC with unit-length capacitors and a passive FIR filter," *IEEE Journal of Solid-State Circuits*, vol. 54, no. 3, pp. 636–645, March 2019.

[137] S. Y. Son et al., "Strain induced changes in gate leakage current and dielectric constant of nitrided Hf-silicate dielectric silicon MOS capacitors," *App. Physics Letters*, vol. 110, July 2011.

[138] Ambiq apollo ultra-low power micrcontroller, product homepage, https://ambiq.com/apollo2/ [online], accessed on July 15, 2021.

[139] M. Elsobky et al., "Relative humidity sensors for system-in-foil applications," *Proceedings of the MikroSystemTechnik Congress*, Munich, pp. 1–3, 2017.

[140] M. Nagata, J. Nagai, K. Hijikata, T. Morie, and A. Iwata, "Physical design guides for substrate noise reduction in CMOS digital circuits," *IEEE Journal of Solid-State Circuits*, vol. 36, no. 3, pp. 539–549, March 2001.

[141] M. M. Ghanbari, J. M. Tsai, A. Nirmalathas, R. Muller, and S. Gambini, "An energy-efficient miniaturized intracranial pressure monitoring

system," *IEEE Journal of Solid-State Circuits*, vol. 52, no. 3, pp. 720–734, March 2017.

[142] American Semiconductor Homepage. [online]. www.americansemi.com /flex-ics.html.

[143] H. Marien, M. S. J. Steyaert, E. van Veenendaal, and P. Heremans, "A fully integrated ΔΣ ADC in organic thin-film transistor technology on flexible plastic foil," *IEEE Journal of Solid-State Circuits*, vol. 46, no. 1, pp. 276–284, January 2011.

[144] C. Garripoli et al., "15.3 an a-IGZO asynchronous delta-sigma modulator on foil achieving up to 43 dB SNR and 40 dB SNDR in 300 Hz bandwidth," *2017 IEEE International Solid-State Circuits Conference (ISSCC)*, pp. 260–261, 2017. doi: 10.1109/ ISSCC.2017.7870360.

[145] A. Grzesiak, R. Becker, and A. Verl, "The bionic handling assistant, a success story of additive manufacturing," *Assembly Automation*, vol. 31, pp. 329–333, 2011.

[146] IMS CHIPS projects overview including kosif project, https://www.ims-chips.de/?page_id=392 [online], accessed on July 15, 2021.

[147] C. Harendt et al., "Hybrid system in foil (HYSiF) exploiting ultra-thin flexible chips," *Proceedings of the European Solid State Device Research Conference*, pp.210–213, 2014.

[148] Product Website. [online]. http://www.ti.com/product/CC1352R, accessed on October 4, 2019.

[149] Product Website. [online]. https://ams.com/as39513, accessed on October 4, 2019.

[150] H. Reinisch et al., "A multifrequency passive sensing tag with on-chip temperature sensor and off-chip sensor interface using EPC HF and UHF RFID technology," *IEEE Journal of Solid-State Circuits*, vol. 46, no. 12, pp. 3075–3088, December 2011.

[151] C. J. Deepu, X. Zhang, W. Liew, D. L. T. Wong, and Y. Lian, "An ECG-on-chip with 535 nw/channel integrated lossless data compressor for wireless sensors," *IEEE Journal of Solid-State Circuits*, vol. 49, no. 11, pp. 2435–2448, November 2014.

[152] Product Website. [online]. https://www.infineon.com/cms/en/prod uct/security-smart-card-solutions/optiga-embedded-security-solu tions, accessed on October 4, 2019.

[153] Product Website. [online]. https://www.emmicroelectronic.com/expert ise/communication/rfid, accessed on October 4, 2019.

[154] Product Website. [online]. https://ams.com/sl900a, accessed on October 4, 2019.

[155] T. T. Nguyen et al., "A flexible bimodal sensor arrays for simultaneous sensing of pressure and temperature," *Advance Materials*, vol. 26, pp. 796-804, Febraury 2014.

[156] J. Wolf et al., "Ultra-thin silicon chips in flexible microsystems," *Proceedings of the 13th Electronic Circuits World Convention ECWC*, Nuremberg, May 2014.

[157] J. Boudaden et al., "Polyimide-based capacitive humidity sensor," *Sensors*, vol. 18, no. 2, 2018.

[158] A. Rivadeneyra et al., "Printed electrodes structures as capacitive humidity sensors: A comparison," *Sensors and Actuators A: Physical*, vol. 244, pp. 56–65, June 2016.

[159] Z. Zhu, G. Yang, R. Li, and T. Pan, "Photopatternable PEDOT:PSS/PEG hybrid thin film with moisture stability and sensitivity," *Microsystems & Nanoengineering*, vol. 3, 2017.

[160] R. M. Morais, M. d. S. Klem, G. L. Nogueira, T. C. Gomes, and N. Alves, "Low cost humidity sensor based on PANI/PEDOT:PSS printed on paper," *IEEE Sensors Journal*, vol. 18, no. 7, pp. 2647–2651, April, 2018.

[161] X. Leng, W. Li, D. Luo, and F. Wang, "Differential structure with graphene oxide for both humidity and temperature sensing," *IEEE Sensors Journal*, vol. 17, no. 14, pp. 4357–4364, July, 2017.

[162] S. Ghosh, R. Ghosh, P. K. Guha, and T. K. Bhattacharyya, "Humidity sensor based on high proton conductivity of graphene oxide," *IEEE Transactions on Nanotechnology*, vol. 14, no. 5, pp. 931–937, September 2015.

[163] H. Bi et al., "Ultrahigh humidity sensitivity of graphene oxide," *Scientific Reports*, vol. 3, September 2013.

[164] H. Guo et al., "Transparent, flexible, and stretchable WS2 based humidity sensors for electronic skin," *Nanoscale*, vol. 9, pp. 6246–6253, 2017.

[165] J. Qian et al., "Positive impedance humidity sensors via single-component materials," *Scientific Reports*, vol. 6, May 2016.

[166] D. Phan, I. Park, A. Park, C. Park, and K. Jeon, "Black P/graphene hybrid: A fast response humidity sensor with good reversibility and stability," *Scientific Reports*, vol. 7, September 2017.

[167] U. Mogera, A. A. Sagade, J. S. George, and U. G. Kulkarni, "Ultrafast response humidity sensor using supramolecular nanofibre and its application in monitoring breath humidity and flow," *Scientific Reports*, vol. 4, February 2014.

[168] M. U. Khan, G. Hassan, and J. Bae, "Bio-compatible organic humidity sensor based on natural inner egg shell membrane with multilayer

crosslinked fiber structure," *Scientific Reports*, vol. 9, 5824 (2019), April 2019. https://doi.org/10.1038/s41598-019-42337-0.

[169] J. Rausch, L. Salun, S. Grieheimer, and R. W. M. Ibis, "Printed piezo-resistive strain sensors for monitoring of light-weight structures," *Proceedings of SENSOR+TEST Conferences*, pp. 216–221, 2011. doi: https://doi.org/10.5162/sensor11/b1.3.

[170] A. Dionisi, M. Borghetti, E. Sardini, and M. Serpelloni, "Biocompatible inkjet resistive sensors for biomedical applications," *Proceedings of the IEEE International Instrumentation and Measurement Technology Conference*, pp. 1629–1633, 2014. doi: https://doi.org/10.1109/I2MTC .2014.6861021.

[171] D. Raiteri, P. v. Lieshout, A. v. Roermund, and E. Cantatore, "An organic VCO-based ADC for quasi-static signals achieving 1LSB INL at 6b resolution," *2013 IEEE International Solid-State Circuits Conference Digest of Technical Papers*, pp. 108–109, 2013. doi: https://doi.org/10 .1109/ISSCC.2013.6487658.

Cambridge Elements ≡

Flexible and Large-Area Electronics

Ravinder Dahiya

University of Glasgow

Ravinder Dahiya is Professor of Electronic and Nanoengineering, and an EPSRC Fellow, at the University of Glasgow. He is a Distinguished Lecturer of the IEEE Sensors Council, and serves on the Editorial Boards of the *Scientific Reports, IEEE Sensors Journal and IEEE Transactions on Robotics*. He is an expert in the field of flexible and bendable electronics and electronic skin.

Luigi G. Occhipinti

University of Cambridge

Luigi G. Occhipinti is Director of Research at the University of Cambridge, Engineering Department, and Deputy Director and COO of the Cambridge Graphene Centre. He is Founder and CEO at Cambridge Innovation Technologies Consulting Limited, providing research and innovation within both the health care and medical fields. He is a recognised expert in printed, organic, and large-area electronics and integrated smart systems with over 20 years' experience in the semiconductor industry, and is a former R&D Senior Group Manager and Programs Director at STMicroelectronics.

About the Series

This innovative series provides authoritative coverage of the state of the art in bendable and large-area electronics. Specific Elements provide in-depth coverage of key technologies, materials and techniques for the design and manufacturing of flexible electronic circuits and systems, as well as cutting-edge insights into emerging real-world applications. This series is a dynamic reference resource for graduate students, researchers, and practitioners in electrical engineering, physics, chemistry and materials.

Cambridge Elements $\equiv$

Flexible and Large-Area Electronics

Printed in the United States
by Baker & Taylor Publisher Services